Synoptic-Dynamic Meteorology Lab Manual

VISUAL EXERCISES TO COMPLEMENT
MIDLATITUDE SYNOPTIC METEOROLOGY

Gary M. Lackmann, Brian E. Mapes *&* **Kevin R. Tyle**

AMERICAN METEOROLOGICAL SOCIETY

Reprinted with corrections August 2024.

Published by the American Meteorological Society
45 Beacon Street, Boston, Massachusetts 02108

The mission of the American Meteorological Society is to advance the atmospheric and related sciences, technologies, applications, and services for the benefit of society. Founded in 1919, the AMS has a membership of more than 13,000 and represents the premier scientific and professional society serving the atmospheric and related sciences. Additional information regarding society activities and membership can be found at www.ametsoc.org

ISBN: 978-1-878220-26-4
eISBN: 978-1-935704-65-2

Front cover image: Three-dimensional representation of output from a numerical weather prediction model, valid 0600 UTC 8 November, 2014, during the extratropical transition of Supertyphoon Nuri in the Western North Pacific Ocean. Contours (mostly obscured) are sea level pressure (4 hPa interval); red isosurface is heating rate from model microphysics; brownish-red isosurface is heating rate from convective parameterization; grey isosurface is heating from planetary boundary layer scheme; purple isosurface is heating rate from radiation; blue isosurface corresponds to 1.5 PVU surface, plotted only below 400 hPa; orange isosurface represents the difference in geopotential height between a full-physics and an adiabatic simulation. These orange areas correspond to locations where the upper-tropospheric height field is higher in the full-physics simulation, consistent with locations above or downstream from regions of heating. The plot illustrates a sequence of downstream baroclinic waves, in this case likely amplified by the recurving and transitioning Nuri, which was at this time located near Alaska. The plot demonstrates the variety of useful diagnostic information available from NWP models, allowing users to see which regions are most strongly affected by different model physics components. A related lesson on the diagnosis of different physical processes accompanying a cyclone is provided in chapter 5.4.

CONTENTS

INTRODUCTION: ORIENTATION AND SOFTWARE

Outline of Chapter:

About This Manual

This laboratory manual is organized in parallel to the *Midlatitude Synoptic Meteorology* (MSM; Lackmann 2011) textbook, with lessons and activities for many of the main chapters of the text. For brevity, the activities in this manual will repeat only a little of the MSM material, but it will be assumed that students undertaking the lab exercises are familiar with the background concepts, drawn from MSM or elsewhere.

Some of the lessons in this manual are *on-paper exercises*, involving only sketching, writing answers and explanations, or writing brief mathematical derivations. Others are *data exercises*, which must be done on a computer.

Software for these data exercises must perform a diverse set of tasks, ranging from multi-source data access (e.g., from directories on a single computer or a local area network, and from remotely accessible data catalogs), to meteorology-specific displays (e.g., horizontal "plan-view" maps and RAOB soundings), to remotely sensed imagery (e.g., satellite and radar displays), to analysis and 3D visualization of gridded output from numerical models. For these reasons, we chose to base the exercises mainly on the *Integrated Data Viewer* (IDV), developed and maintained by Unidata. Unidata is one of the programs within the non-profit University Corporation for Atmospheric Research, funded since the 1980s by the U.S. National Science Foundation to "democratize" access to weather data. With that long history, their software is very mature and well-designed, and remains actively supported by both dedicated personnel at Unidata and the open-source software developer community worldwide.

The IDV is a free, easy-to-install, open-source application. Written in the Java programming language, it runs on all contemporary computer operating systems. It reads from (and can write to) data files and catalogs stored on remote servers, or locally on a user's computer. It features a powerful array of diagnostic and visualization capabilities, including meteorology-specific special displays. Displays in IDV are inherently three-dimensional, although we often use simple "plan views" (on a flattened Earth map, viewed from above). Of course the vertical coordinate is greatly exaggerated in 3D synoptic displays: the troposphere is really thinner than an onion's skin, in a relative sense. The IDV remains under active development by Unidata staff, with input from its user community (which includes the authors and users of this manual!), so please feel free to send us (and/or Unidata) suggestions and feedback.

Each exercise in this manual uses these typefaces for clarity:

Normal typeface is used for background information, technical instructions, motivating questions, and learning objectives. **Bold indicates assigned actions and questions that students are expected to respond to in their report.** A `constant width` typeface is used to indicate text that can be found exactly on the IDV software (usually on the `Dashboard`, `Map`, or `Globe` view windows, and display `Legend` areas).

The word **Optional:** is used to set off suggestions for further explorations, which may be conceptual, or involve alternative data sources. For instance, in some exercises, the IDV can be dialed back to arbitrary dates to examine past archival cases, or real-time data can be examined with lookalike displays identical to the lesson's example.

This typeface convention is repeated on the introductory page of each chapter.

Introduction to the Integrated Data Viewer (IDV)

To use this manual, you will need to install and test this free software, and install a few plug-ins for additional functionality. The main Unidata IDV web page* gives its overview. The IDV is open-source, free software; its installation involves just a few clicks and a minimal registration process. The download link is easily found on the main IDV web page referenced above. Students should be encouraged to install it on their own computers as well. The IDV is memory intensive: to perform well, a computer should have at least 4GB of RAM, preferably more (especially given the relatively high memory demands of current computer operating systems). Other large applications and multiple web browser tabs and/or windows should be closed to conserve memory. Efficiency tweaks in recent IDV versions have made this less of a problem, happily.

Please refer to Appendix 1 of this manual to orient your expectations. The IDV is professional scientific software, which we liken to a workshop full of sharp and powerful tools, often without blade guards. Many pitfalls and frustrations await a careless approach to such a workshop! In the age of commercial software, students may be accustomed to poking at buttons thoughtlessly or impatiently, a habit that must be discouraged in this case. More information is also available at these sources:

IDV reference manual (also available under the IDV's `Help` menu)
http://www.unidata.ucar.edu/software/idv/docs/userguide/toc.html

Tutorial from the IDV training workshop:
http://www.unidata.ucar.edu/software/idv/docs/workshop/

---

* http://www.unidata.ucar.edu/software/idv/

Tutorial screencasts on YouTube:
https://www.youtube.com/user/unidatanews/

Appendix 1 also introduces the *Mapes IDV collection,* a free curated set of resources aimed at making the IDV more powerful, yet more usable. You install the *Mapes IDV collection* as a *Plugin,* an extension to Unidata's generic IDV distribution, as shown in Appendix 1 and in the next section (**0.1**). The *Mapes IDV collection* has a prominent section devoted to `LMT manual materials`. These materials are all accessible, free of charge, from any machine on the Internet. Users are invited to explore the many other aspects of the collection as well.

0.1. Install, Launch, Install Plugins, and Relaunch

INSTALL

As mentioned above, you can download the IDV from this link: http://www.unidata.ucar.edu/software/idv/. After registering with Unidata, download the appropriate version for your operating system. Choose the **Current Release** of the software (presently 5.4). The *64-bit* version is default for Macs, and is strongly recommended for Windows and Linux, unless you know for sure that your computer is only running a 32-bit version of these two operating systems (note that the memory footprint of the 32-bit version of IDV is limited *to no greater than 1.5 GB of RAM,* which may be insufficient for some of the exercises in subsequent chapters of this manual). Follow the prompts to install the IDV; the default options are recommended unless you really know what you are doing (if you are installing on a Linux machine, you may need to work with your system administrator in order to complete the installation).

LAUNCH

Once you have the IDV installed, launch it. To do this on a Windows machine, double-click on the IDV icon which is by default placed on the Desktop during installation. You will likely be prompted to allow the IDV to access network ports when it first loads; assuming you have Administrator privileges on the computer, go ahead and authorize its use on all network interfaces (see screenshot below).

Mac users will find IDV in its own subfolder within Applications. Go into that folder and double-click on the IDV icon. If desired, move it to the Dock or Desktop for easier access. Linux is a bit more complicated; the recommended use is to note the directory path where the IDV is installed; then open up a command window, **cd** into that directory, and type

./runIDV to launch the IDV (for savvy Mac OS users, a similar command-line technique can also be used within that operating system's **Terminal** application).

INSTALL PLUGINS

We recommend the following *Plugins* to improve the bare IDV installation (note that these are separate from the *Mapes IDV Collection* Plugin that you will install separately). To enable these, find the menu item `Tools > Plugin Manager` in the menu bar of either the `Dashboard` or `Map View` windows. That will pop up the `Plugin Manager` window, where you will see many choices:

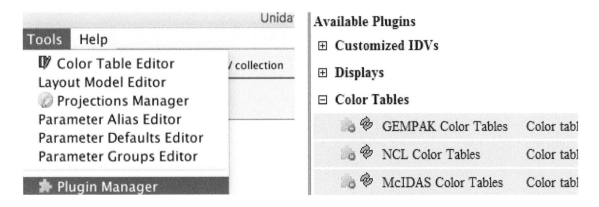

Click the green button with the white cross to the left of each Plugin you want. If you plan to install several, select `No` in the `Plugin Confirmation` window: You can restart later, after adding them all.

Recommended plugins:

- **The Mapes IDV collection (under Miscellaneous)**
- **Color Tables**: Install any or all the color tables offered.
- **Maps**: Consider *US Highways and Roads* and *US Rivers & Lakes*
- **Bug Fixes**: Install the *Spurious Map Lines Fix* Plugin. This is often necessary on computers whose graphics cards are based on the AMD/Radeon chipsets.

The next time you launch IDV, these Plugins will be present. For instance, you can check if the color bars you added are present by navigating to the menu item `Tools > Color Table Editor`, and examining its `Color Tables` menu like this:

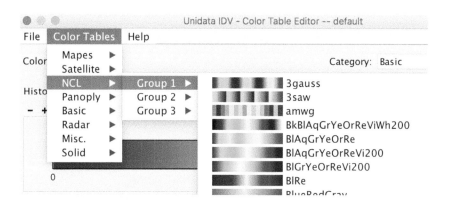

All the above Plugins are *static files*. They are now stored as unchanging files in the user's home directory, under the $HOME/.unidata/idv/DefaultIdv/ subdirectory on Linux and MacOS, or C:\Users\<user name>\.unidata\ subdirectory on Windows. This directory, which can be thought of as an *IDV Preferences* folder, gets created when the user first launches the IDV.

RELAUNCH

You must quit and relaunch the IDV for your Plugins to take effect. In addition, you must have an active internet connection for the dynamic resources of the *Mapes IDV collection* Plugin to function properly. If you are offline, the IDV will still launch, but without the *Mapes IDV* resources including bundle and data catalogs and other features mentioned in this manual.

It is possible to work offline, however. IDV bundle files with the `.zidv` suffix—the heart of most of the data exercises in this manual—can be downloaded ahead of time and made available as needed on classroom computers; the most current versions can be down-loaded at any time via the web, from the IDV Bundles folder tree of the Mapes repository at http://bit.ly/Mapes_IDV. Any of these can be opened in the IDV with the `File > Open` menu, allowing one to work without an internet connection. You can download the entire current set of IDV bundles comprising the **LMT manual materials,** using your browser, from the menu shown below:

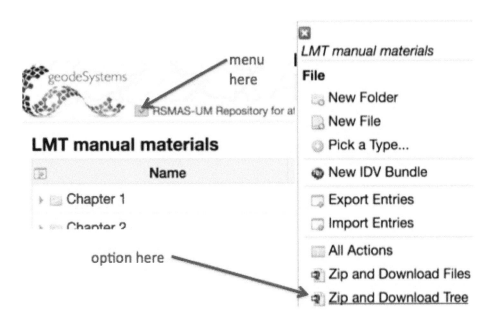

0.2. Test-Drive Using a Pre-made IDV Bundle

The IDV application saves its entire state in an XML file called a "bundle," with suffix *.xidv* or *.zidv*. The z in *.zidv* indicates a zipped file with the actual data included, allowing offline use on any machine. Opening a bundle file (for example, with `File > Open` in the menus) will restore the IDV's state that was present when the bundle file was created, including all data sources, displays, and even the particular 3D viewpoint.

Let's illustrate basic IDV use with one of the predefined *.zidv* bundles from the manual.

OPEN AN EXAMPLE BUNDLE FROM CHAPTER 1

As a test drive of the software, this section walks you through a set of actions that will get you familiar with the IDV software. The scientific lessons will be deferred to Chapter 1.

Launch the IDV. You will see two main, separate windows: a `Dashboard` and a `Map View` window. If you haven't disabled it, an `IDV Help Window` will also appear. If you installed the *Mapes IDV collection* Plugin correctly, and then you launch IDV while online, your `Toolbar` area on both windows will contain the Mapes IDV collection favorites folder, like this:

To open the example bundle for this section, find it in the repository:

Find LMT bundles: approach 1:

Left-click the `Mapes IDV collection folder.` Menus will unfold to reveal `Classes and Labs > LMT manual materials > Chapter 1 > LMT_1.2.` Select this item.

Find LMT bundles: approach 2:

You can find the same folder tree in the `Catalog` area of `Data Choosers` tab of the `Dashboard`. Select `General > Catalogs` in the sidebar at left, and drill down into the `Mapes IDV collection` folder to find LMT_1.2, as shown in the screen capture below. To launch the bundle, double click it, or highlight it and click the `Add Source` button at the bottom of the `Dashboard` window.

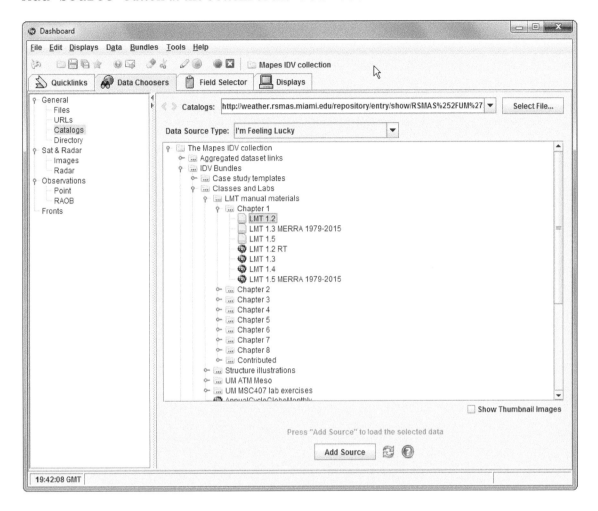

After either approach:

Once you have launched the bundle, accept the default suggestions in the `Open bundle` and `Zip file data` dialog windows shown below.

Do *not* check the `Don't show this again` box. Some exercises involving successive additions of bundles of displays will become impossible. If this is done inadvertently, it can be restored by clicking on `Edit > Preferences`; look under the `General` properties tab:

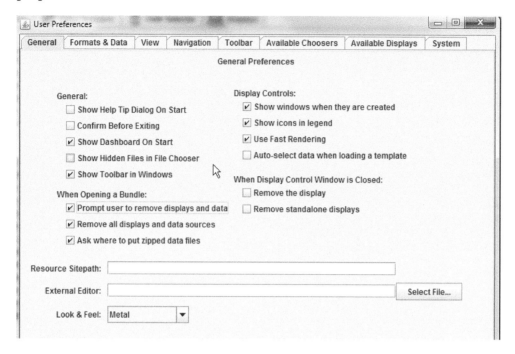

In a few moments, your `Map View` window will have a `Legend` on its right-hand side, with many displays indicated, similar to that shown below.

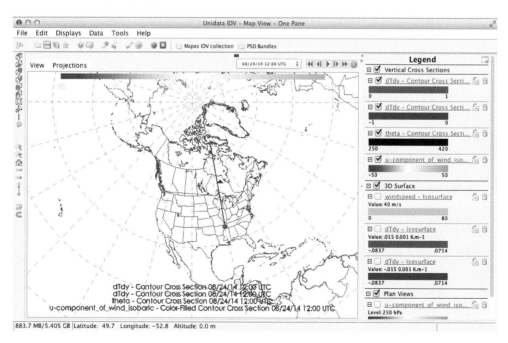

As a brief tour of the possibilities that this IDV bundle presents to you, try the following actions.

A 3-button mouse with scroll wheel is recommended to access the IDV's rich suite of features. However, some of the same functionality exists in two-fingered swipes on trackpads, with or without the `control` key depressed, or using arrow keys. Under the `Help` menu, you will see a special section on panning, zooming and rotating.

- Under `3D Surface,` check the box for the green `windspeed - Isosurface` display. The jet stream becomes visible.
- Click and hold the right mouse button over the map display, and move the mouse around. You will see 3D motion of the display volume, revealing the cross sections. (Left and right arrow keys or two-fingered gestures with and without `control` will perform the same action.)
- Repeat the above mouse action with the `control` key depressed. This time the display volume will pan rather than rotate. Alternately, green arrows at the left edge of the view window will pan the view manually.
- Zoom in and out with the mouse scroll wheel, or use the magnifying glass icon ⌕. A vertical two-finger gesture on a trackpad also does this.
- Jump to a west view by clicking this blue-sided cube icon ⬚ on the left edge of the window. Use 3D rotation (holding the right mouse button while moving the mouse) to see how this west view fits into the 3D scene.
- Use the blue-topped cube ⬚ or Home 🏠 icon to return to top view.
- In the Legend, minimize or expand the details of various displays, or *groups* of displays, with the colored ⊟ and ⊞ squares to the left of the visibility checkboxes.

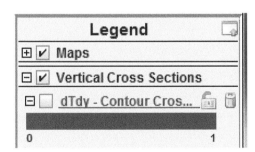

- A display can be eliminated entirely with the trashcan 🗑 icon.

Additional basic operations for interacting with the view and the displays are also accessible using the icons along the left-hand edge of the window. Explore them. These and many other aspects of the IDV are documented under the `Help` menu.

A WINDOWS OPERATING SYSTEM-SPECIFIC MISBEHAVIOR

You may occasionally notice that your IDV display window goes "blank" if on a computer running the Windows operating system:

Should this occur, try left-clicking in the blank region, or simply resize the window slightly by clicking anywhere on the outer periphery of the `Map View` pane and drag to slightly resize. The display should reappear.

ADD A NEW DISPLAY

- Create a Blue Marble image, by clicking the ⊕ icon in the toolbar. Notice that the new display appears under `Maps` in the `Legend` of the `Map View` window. Expand the `Maps` menu if necessary by clicking on the plus sign to its left, and then use the checkbox to turn its visibility on or off, whichever you prefer.
- To add a more advanced display, locate the `Dashboard` window. If you have lost or closed it, re-open or create a fresh one by clicking on this gamepad 🎮 icon in the upper left corner of the `Map View` window. Click the `Field Selector` tab near the top of the `Dashboard` window. You will see `Data Sources` listed along the left edge of the window. Select `GFS CONUS 95km`. Click the unpacking icon next to `3D grid` (a small triangle or knob-like symbol), successively, until you are able to select `Temperature @ Isobaric surface`.

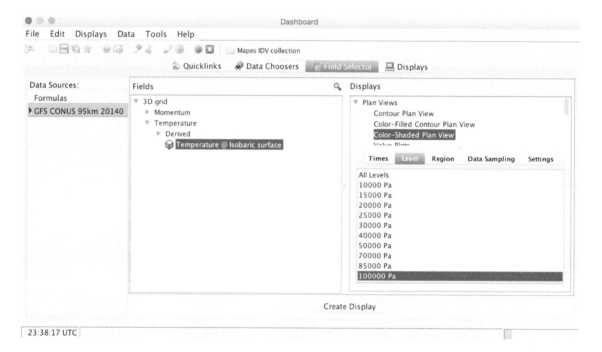

At that stage, choices appear in the `Displays` area at right. Try the following:

- `Displays selector`: Select `Color-Shaded Plan View` under the `Plan Views` tab, a safe choice (contours have many options which can make a display fail to be visible).
- `Times` tab: there is only one time in this case.

- `Level` tab: choose 100000 Pa (or 1000 hPa; this is near sea level).
- `Region` tab: choose `Match Display Region` in the dropdown menu.
- `Data Sampling` tab: take no action to accept the default (full resolution).
- `Settings` tab: take no action.
- `Create Display` is the button at the bottom of the `Dashboard`. Click it.

Your `Map View` window will now have a new `Temperature isobaric` display, as shown below.

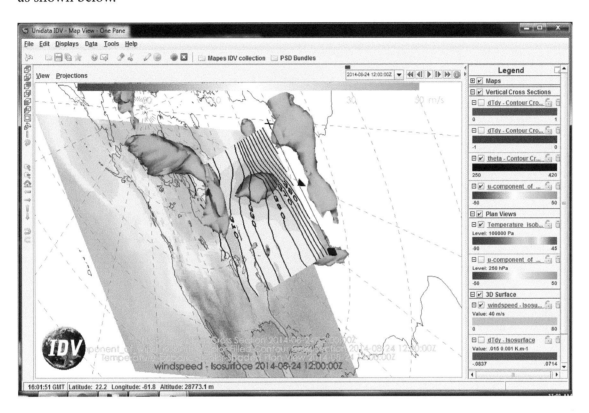

In the `Legend`, find your new display called `Temperature_isobaric`, under `Plan Views`. Confirm it by unchecking and rechecking its visibility checkbox. Now click the blue hyperlink itself (<u>Temperature isobaric...</u>). A `Display Control` window will appear, either as a tab in the `Dashboard` window if you still have that open, or otherwise as a standalone window.

Try the following:

- `Color Table:` Click the `Temperature` button to choose a new color table. You can also try changing the range of the color table, as the default range of −90 to +45 is rather wide.
- `Smoothing:` Apply a horizontal smoothing, such as `Gaussian weighted,` and notice how the display changes.

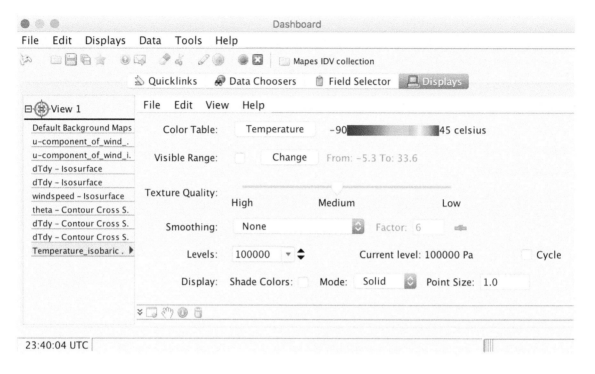

- `Levels:` Change to 25000 (the 250-hPa level, typically in the upper troposphere). In 3D views, you will see that this is close to the level of the mean jet stream.
- On the `Map View` window's Legend, check the box to make the green `windspeed-Isosurface` display of the jet stream visible. Now rotate the display in 3D by holding down the right mouse button as you move the mouse, to verify that your 250 hPa Temperature display plane is at about the same altitude as the jet stream.
- Add a Color Scale. To do this, from the `Edit` menu of the `Display Control` window, select `Properties...` A new window will pop up. One of its tabs is `Color Scale`. Select `Visible` and adjust as desired, then click `Apply` or `OK`:

- Capture an image or movie: In the `Map View` window's display area, find the `View` menu. Select `Capture > Image`. Set the file name suffix to determine the graphical image type, such as .jpg or .png.

SAVE AND RESTORE YOUR SESSION

- Save your IDV session state as a bundle file called `test.xidv`, using the `File > Save As...` menu, or the floppy disk icon ⊟ in the toolbar.
- Erase your displays and data with the scissors icon, or only the displays using the eraser ✐ ⚒ icon (mouse-hover will show what the icons do). Or, quit and restart the IDV. Quit the IDV by either clicking on the ⊠ icon; selecting `File > Exit`; or simply by closing the `Map View` window itself. Restart IDV just as you started it before.
- Now, open test.xidv from the `File > Open` menu. Or, find it and other recently used files in the `Dashboard's Quicklinks` top tab, in the `History` sub-tab on the left hand side.

Your session should appear exactly as it was when you saved it. (Note: In this case, this only works because the data are still in the temporary directory they were unzipped into.)

For additional IDV skill building, with detailed step-by-step screenshot instructions like the above, try Exercise 3.4.

0.3. Key Points of This Chapter

We hope this tour has shown you the professional, "workshop"-like usage style of the IDV. More advice on working with the IDV is expressed in the orientation discussion and linked materials in Appendix 1.

You can now see why using collections of pre-built bundles of displays is so much easier than the arduous process of creating your own displays from scratch. Perhaps the exercise showed you the need for caution too. For instance, when you changed the `Color Table` on your display, there was no "Back" or "Undo" button, a common feature in commercial software packages such as Microsoft Office or Photoshop! If you had created a carefully customized color table, and then you changed it, all that prior effort would be lost. *If you don't save your bundle, manually, every time you add valuable effort, your work can be lost with a slip of the mouse.* The wry advice of an author's wood and metal shop teacher applies: "Measure twice, cut once"—just as you would with real workshop tools. You can surely see how students (and instructors!) accustomed to operating mass-market commercial software and apps might find the IDV frustrating at first.

With this cautionary advice in mind, we hope you will agree that IDV's power in making 3D meteorological visualizations is worth the care and effort required to teach and learn well with it.

Appendix 1: Further IDV Orientation Discussion and Materials

A PowerPoint presentation with further orientation material is available at http://weather.rsmas.miami.edu/repository/entry/show?entryid=dd95b65c-09a5-43a5-9f44-da5243e302f4 or its shortened link http://bit.ly/2m9zS9s.

Appendix 2: Case Suggestions for Historical Storm Analyses

In some **Optional:** exercises, displays identical to those in the exercise can readily be generated for use on historical cases, using free online archival data sources. Bundles with the form LMT_X.Y_MERRA_1979-2015 mark such opportunities.

Below is a very partial and arbitrary listing of past storm cases that readers could view:

Over the United States:

- Inauguration Day storm of 1993 (Pacific Northwest, 20 January 1993)
- "Storm of the Century" https://en.wikipedia.org/wiki/1993_Storm_of_the_Century (12–15 March 1993)
- The 10 November 1998 storm examined in detail in Chapter 8 of the excellent textbook of Wallace and Hobbs (2006); utilized in exercise 5.3
- Superstorm Sandy (October 2012)

Beyond the United States:

- European storm Xynthia (https://en.wikipedia.org/wiki/Cyclone_Xynthia, 27 Feb–1 Mar 2010)
- The extratropical transition (ET) event following STY Nuri in the western Pacific https://en.wikipedia.org/wiki/Typhoon_Nuri_(2014) (20 October–10 November 2014)

Many more cases of interest could be found, for instance in Asia or the Southern Hemisphere. Cases that cross the dateline will unfortunately not be possible. If a community of readers takes the time to create high-quality case studies, we will gladly publish (host) those files in a repository. To contribute, save your analysis bundle as a zipped (.zidv) file, and contact mapes (at) miami.edu for options to add it to the repository. A folder called Contributed has been created in the LMT materials area, in preparation for this hoped-for outcome.

1

BASICS

This chapter includes the following exercises:

1.1. Thermal Wind and Temperature Advection
1.2. Thermal Wind and the Jet Stream
1.3. Thermal Wind and Temperature Advection in a Winter Storm
1.4. Calculation of the Vertical Component of Relative Vorticity
1.5. Rossby Wave Excitation by a Recurving Typhoon

Each exercise in this manual uses these typefaces for clarity:

Normal typeface is used for background information, technical instructions, motivating questions, and learning objectives. **Bold indicates assigned actions and questions that students are expected to respond to in their report.** A constant width typeface is used to indicate text that can be found exactly on the IDV software (usually on the Dashboard or Legend areas).

The word **Optional:** is used to set off suggestions for further explorations.

1.1. Thermal Wind and Temperature Advection

Revisit section 1.4 in *Midlatitude Synoptic Meteorology* (MSM) for background information corresponding to this lesson.

Students of atmospheric dynamics know that in the Northern Hemisphere, a *clockwise* turning of the geostrophic wind with height (a *veering* wind profile) is associated with warm advection in the layer over which the geostrophic shear is evaluated. Similarly, *counter-clockwise* turning of the geostrophic wind with height (a *backing* wind profile) corresponds to cold advection. Can you recall why? The objectives of this short exercise are (i) to illustrate why and how turning of the geostrophic wind with height relates to temperature advection in the layer, and (ii) to evaluate the relation between the thermal wind and the *thickness*, which is proportional to the mean layer virtual temperature.

Consider Fig. 1.1 on the following page, which shows a highly idealized set of 1000- and 500-hPa geopotential height isopleths for a hypothetical Northern Hemisphere location. The dashed contours represent 500-hPa height, and solid lines represent 1000-hPa height.

a) **Examine the height contours and, based on your knowledge of the geostrophic wind, draw straight vector arrows at location A to indicate the 1000- and 500-hPa geostrophic wind.** Take some care to estimate the relative magnitudes of these vectors. Ordinarily, the 500-hPa height gradient is greater in magnitude than that at the 1000-hPa level, but that is not the case in this simplified example. **In what sense (clockwise or counterclockwise) is the geostrophic wind turning with increasing height in this layer? What term do we apply to describe this type of vertical turning of the wind with height (veering or backing)?**

b) **On the diagram, use "graphical subtraction" (explained below, or see section 1.4.2 in the MSM text) to analyze the 516-, 522-, 528-, 534-, 540-, and 546-dam 1000–500-hPa thickness isopleths.** To do graphical subtraction, for each thickness value, find at least two intersection points of the given curves where the difference (thickness) has this value. **Now connect these points with a smooth dash-dotted line or curve (a contour of the thickness field). Given that thickness contours essentially represent layer-average virtual temperature isotherms, indicate the location of relatively warmer and colder air on this diagram.**

c) Recall the mathematical expression for the temperature advection by the geostrophic wind, and its meaning in terms of the cross-contour component of vectors. Considering the vectors you drew in a) and the thickness contours from b), **answer the following:**

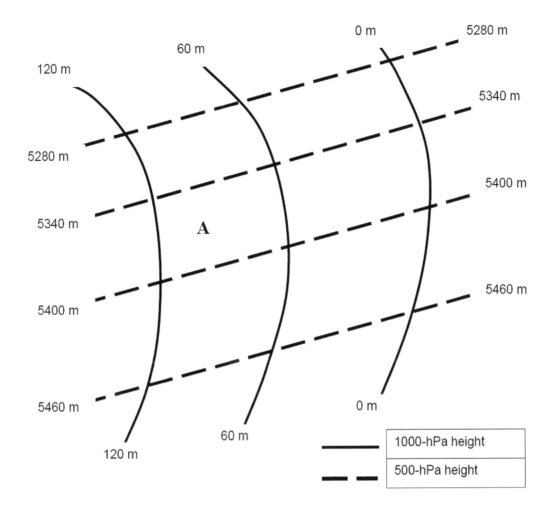

Figure 1.1. Idealized height pattern for a hypothetical Northern Hemisphere location. The dashed contours represent 500-hPa geopotential height, and solid lines represent 1000-hPa height.

i. **If the temperature gradient is uniform through the 1000–500-hPa layer, what is the sense (warm or cold) of the *geostrophic temperature advection* at point A at the *1000-hPa* level. How about at the *500-hPa* level?**

ii. **Estimate the sense of temperature advection in the *middle* of the layer, for example near the 700-hPa level. State any assumptions used in this estimation.**

iii. **Use vector subtraction to draw the thermal wind vector at point A.** See Eq. (1.38) and Fig. 1.8 in the MSM text if you need to review. What is the "rule" regarding the orientation between the thermal wind vector with respect to the thickness contours? **Are the vectors you've drawn consistent with this rule? If not, resize your geostrophic wind vectors from a) to maintain consistency with the required relation.**

d) Think about the relation between the height contours and the thickness (temperature).

Then, think about the relation between the geostrophic wind and the height contours. Now, **explain, in your own words,** why a backing geostrophic wind profile in the Northern Hemisphere guarantees geostrophic cold advection, and conversely why a veering profile implies warm advection.

Optional:

e) On a blank sheet of paper, **sketch** a situation analogous to Fig. 1.1, but with the opposite sign of temperature advection.

f) On a blank sheet of paper, **sketch** a cold advection situation in the *Southern Hemisphere*. Do the Northern Hemisphere "rules" for the relation between veering and backing and temperature advection hold in the Southern Hemisphere?

g) Try to **sketch** a situation for the Northern Hemisphere in which a veering geostrophic wind *does not* correspond to warm advection. Is this possible?

h) Consider the effect of friction on actual lower-tropospheric winds (not *geostrophic* winds). **Do the winds veer or back through a layer with frictional effects decreasing with height?** What are the implications for diagnosing advective tendencies from cloud motions you might observe outdoors, or in the lowest portion of a rawinsonde wind profile (see MSM section 1.4.3)?

i) What sign of temperature advection corresponds to situations when the 1000- and 500-hPa geostrophic wind vectors are oriented *exactly opposite* to one another?

1.2. Thermal Wind and the Jet Stream

The objective of this exercise is to apply the thermal wind concepts to provide an explanation for the structure and position of the *westerly jet stream*. Students will learn to interpret the jet stream's relationship to horizontal temperature gradients, in the context of the thermal wind relation, using gridded data.

a) Load the IDV bundle LMT_1.2.

To do that, just as you did in exercise 0.2, you can start from the Mapes IDV collection folder in IDV's toolbar, following the folder tree to select `Mapes IDV`

```
collection > Classes and Labs > LMT manual materials >
Chapter 1 > LMT_1.2
```
as shown here:

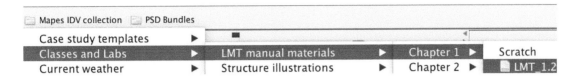

You should see a generally blank North American map with a north–south-oriented cross-section line, as shown in the screenshot in the introduction. Hold the right button as you move the mouse to the right, to rotate the display so that you can view the plane of the cross section from the west. For a direct western side view, click the ⌗ icon in the viewpoints toolbar along the left edge of the window. In the "vertical cross sections" area of the legend, make sure that both the "u-component" and "theta…" displays are activated (checked), so that you see both the wind speed (shaded contours green and purple) and isentropes (contours with labels in K). Locate the zone of strongest westerly flow (green), which we will define as the "jet core" in this cross section. **Describe the slope of the isentropes below the level of the jet core, and compare this to the slope above the jet core. Explain why the isentropic slope reverses as one crosses from beneath to above the jet core.** Hint: Review section 1.4 in the MSM text if needed, and see its Fig. 1.9.

b) In the Legend window, select `windspeed - Isosurface,` which shows you the 3D jet stream core (speed > 40 m s^{-1}). Jumping to a top view for a moment (by clicking either the ⌗ icon or ⌂ icon), it should be more apparent why this particular cross-section location and orientation was selected. **What do you expect the signs of *dT/dy* to be above and below the jet? Explain why this must be the case.** Hint: See Eq. (1.44) in the text.

c) **Now, check your speculation from b).** In the `Legend`, check the boxes for `dTdy - Isosurface,` one for a positive value (red shading) and one for a negative value (blue shading). Rotate the view to see the 3D relationships. Zoom and pan to **isolate an area where the three-layer vertical structure of blue, green, and red isosurfaces is clear. Capture an image** (`View > Capture > Image` or use your computer's screen capture capabilities), and **describe** what you see there. According to the thermal wind relation for the geostrophic zonal wind component, what color of shading corresponds to regions where the westerly wind component is increasing with height? What color corresponds to locations where it is decreasing? **Did the zones of positive and negative meridional temperature gradients** (`dTdy`) **match your expectations from a) above?**

d) In the `Legend`, uncheck the box for `3D surface`, which will deactivate all three isosurfaces in that category, and check the box to visualize all the `vertical cross sections`. Drag the section endpoints to move the cross section to another east–west section of the jet stream, such as south of Alaska. Recall that the thermal wind is directly related to the horizontal temperature gradient, and that the change in the *zonal* (*east–west*) geostrophic wind with height is specifically related to the *north–south* temperature gradient by Eq. (1.44) in the MSM text. Accordingly, we have dT/dy contour cross sections that we can now inspect to see to what extent the real data match the thermal wind relation. **Capture a west view (east-facing image) showing this jet, along with the `theta` and `dTdy` cross section contours, and include this image in your lab write-up.**

e) Consider a meridionally oriented jet stream (with northerly or southerly flow). If you were given a cross section of potential temperature, could you sketch a reasonable set of corresponding isotachs (of the meridional wind distribution)? If you were given the isotachs, could you sketch a reasonable set of isentropes (potential temperature contours)? On a piece of scratch paper, **draw** a blank east–west-oriented cross section, and include a large "J" to mark the jet core. Then **sketch** a few isentropes above, through, and below the jet core. **Label** regions where dT/dx is positive and negative.

f) Suppose that your best friend, who happens to be majoring in underwater basket weaving, looks over your shoulder while you're studying the thermal wind in preparation for an upcoming quiz. Being the curious sort, they ask you "What's the thermal wind?" **Provide a written answer, in three sentences or less**, that would be appropriate given this audience.

g) **Optional:** Examine real-time (RT) weather data in same format. Load the bundle `Mapes IDV collection > Classes and Labs > LMT manual materials > Chapter 1 > LMT_1.2_RT`. Orient the cross sections with respect to a current jet stream, selecting an east–west-oriented jet stream segment (so your cross sections are north–south oriented). **Capture a west view showing the jet velocity, along with the `theta` and `dTdy` cross section contours, and include this image in your lab write-up.**

1.3. Thermal Wind and Temperature Advection in a Winter Storm

Based on section 1.4.2 of MSM and lesson 1.1 in this manual, we recognize that the turning of the geostrophic wind with height is an indicator of warm or cold advection. Here, our objective is to apply these concepts to a winter storm event from December 2009.

Learning outcomes are 1) to evaluate the utility of the thermal wind concept in distinguishing regions of warm and cold temperature advection, and 2) to anticipate how temperature advection regimes impact sensible weather, including clouds and precipitation.

Following a similar procedure to that used in the previous exercise, load the bundle `Mapes IDV collection > Classes and Labs > LMT manual materials > Chapter 1 > LMT_1.3`. You should see 500-hPa height and sea level pressure contours in black and red, respectively. The pink square designates the location of the *skew T vs. log(p)* sounding profile, which appears in its own window.

a) Drag the pink square to a point south of Louisiana, to the west of the surface cyclone center, as shown in the left image below. **Sketch the geostrophic wind directions at low and middle levels**, based on the contours of sea level pressure (`MSLP`, red) and 500-hPa height (`HGHT`, black). Now, **capture a sounding image for your write-up. Does the turning of the wind agree with your geostrophic deductions from the plan-view maps? Is the wind veering or backing in this sounding? What sign of temperature advection is implied?** Check the box to activate the colored 2-m temperature (`t2m`) display. **Does the near-surface geostrophic wind implied by the SLP, acting on the near-surface temperature gradient, agree with the turning of wind direction in terms of indicating the sign of temperature advection? Explain.**

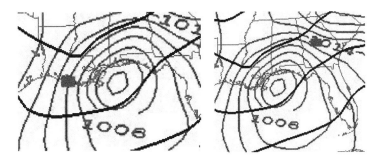

b) Next, move the sounding location into central Georgia, north and east of the cyclone center, as shown in the right image above. **Capture a sounding image for your write-up. Answer the questions from a) above for this new location.**

c) Compare your two soundings from the above steps. **Which of the two indicates a smaller dewpoint depression aloft, and thus more likely indicates thick clouds and precipitation near this location? Explain and justify your answer.**

d) **Using the results above, postulate a "rule" about how the degree of saturation evident in a sounding relates to the sign of the lower-tropospheric temperature advection. Move the sounding around and see how generally your rule applies.** We will return to this in chapter 2, in connection with the quasigeostrophic (QG) omega equation.

e) **Optional:** Discuss how you might use information from the movable sounding probe to help you locate frontal boundaries. Sketch on scratch paper how the horizontal wind field changes with height (pressure) ahead of and behind a cold front. Repeat for a warm front. Using the wind profiles from the soundings, where would we expect the warm and cold fronts to be located in this example?

f) **Optional:** Let's repeat the same displays for another situation, for any time from post-1979 through February 2016. To do this, load the bundle `Mapes IDV collection > Classes and Labs > LMT manual materials > Chapter 1 > LMT_1.3_MERRA_1979-recent`. Initially the same date and time (December 18, 2009) will come up again. If you are using IDV versions before 5.2u1, the wind barbs will not be visible on your sounding window. In this case, follow the screenshot below in order to select `Black on white`.

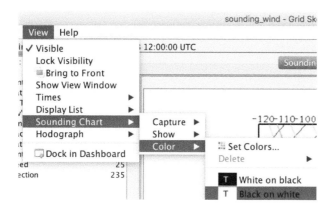

To change the geographical region examined:

- Zoom out if needed so that the map display includes your area of interest.
- While holding down the shift key, hold down the left mouse button as you drag or "rubberband" to define a focus box on the map.

To change the date examined:

- Click the ⓘ icon in the animation control area to call up the `Time Animation Properties` menu.
- Adjust the `Start Time` in the `Define Animation Times` tab of that menu to any date and time between January 1, 1979, and early 2016.
- If you like, adjust also the `End Time` and `Interval` to include several time steps. *Use appropriate care* to limit your time range in light of available computer RAM and long delays in loading too much data from the NASA server. (This is why we recommend adjusting the display region before expanding the time span.)

- If you want to save your case, use the `File > Save As...` dialog from the main display window.

1.4. Calculation of the Vertical Component of Relative Vorticity

Revisit section 1.5 in MSM for the text that corresponds to this lesson.

One of the most commonly utilized variables in atmospheric dynamics is *vorticity*. Frequently, we view displays of vorticity without thinking about the calculation itself, and how the vorticity field is related to the wind field at that moment. In this exercise we will compute vorticity manually, and then compare our manual calculation to a plot of vorticity generated by the IDV. The learning outcomes are 1) to allow students to calculate the vorticity based on analyzed wind values, thereby gaining an appreciation for the "behind the scenes" work in generating plots of vorticity (or similar *kinematic* fields: spatial derivatives of the winds), 2) to consider sources of error and the sensitivity of such calculations to dataset resolution and other factors, and 3) to evaluate the magnitude of vorticity in this case, and discuss its representativeness in the midlatitude atmosphere.

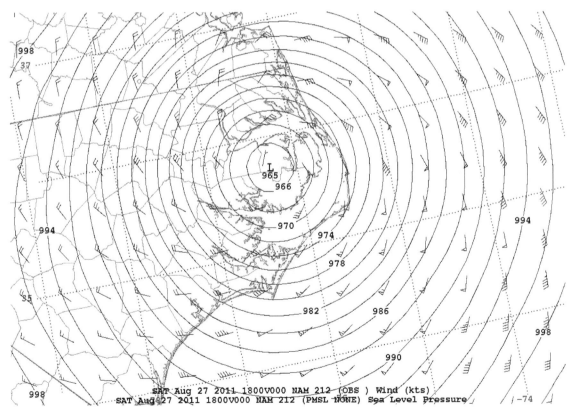

Figure 1.2. NAM model analysis of sea level pressure (black contours, 4-hPa interval), and near-surface wind (barbs) valid at 1800 UTC 27 August 2011. Latitude and longitude lines are shown at 1° intervals as dotted lines.

The image below shows the near-surface wind field and sea-level pressure analysis from the 40-km North American Mesoscale (NAM) model valid at 1800 UTC Saturday 27 August 2011. Latitude and longitude lines are shown at 1° intervals [1° of latitude = 111 km; you can use this distance as a spatial ruler, or compute the spacing between longitude lines as 111 km × cos(latitude)].

a) **Estimate** $\zeta = \partial v / \partial x - \partial u / \partial y$ **for a region near the storm center**, using data in Fig. 1.2. **Show your work.** You will need to use "finite difference" approximations to the derivatives, and estimate the u and v wind components from the barbs shown (1 knot = 0.51 m s^{-1}). Use a distance of 1° of latitude (111 km) to set up a grid of points at which to evaluate u and v wind components, and use this distance in the denominator to compute spatial derivatives. **State any assumptions you use in this computation.**

Value of vorticity near storm center: _____ (include mks units)

b) Now, compare your manual calculation to the result from the IDV software for this same dataset. To do this, open the LMT_1.4 bundle. What range of shaded relative vorticity values is displayed? To find out the units and scale in the `relvort` display, click on its hyperlink in the `Legend` to reveal its display control in the `Dashboard`. Examine the color table, as shown in this screen capture. If you like, change the units with the menu action `Edit > Change Display Unit`.

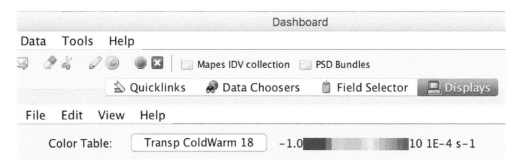

What is the maximum value on the model grid? To read off exact field values, hold the middle mouse button with the cursor in the desired location. Numerical values are shown at the bottom of the window (caution: if your display window is maximized, the readout may be hidden). **How does this compare to a typical midlatitude value of the Coriolis parameter or "planetary vorticity" f?**

c) Experiment with the `Smoothing` options in the `relvort` display control until you think the resolution of the smoothed display matches the 1° scale of your calculation. **Now how well do the values compare? Discuss** at least one reason that could explain the discrepancy.

d) **Optional:** How do the maximum values displayed here compare to those near the surface in other, more typical midlatitude situations? How do the values compare to those typically seen at higher altitudes, such as near the jet stream? In order to answer these questions, **write down a *scale analysis* for the relative vorticity for synoptic-scale, midlatitude conditions**. Based on the results above, would this scaling work for tropical cyclones? If not, what specific scale or scales should differ from those used in midlatitude synoptic weather systems?

1.5. Rossby Wave Excitation by a Recurring Typhoon

Revisit section 1.5.3 in MSM for the text that corresponds to this lesson.

This lesson illustrates how Rossby waves in the upper-level midlatitude jet stream can transmit the impacts of a tropical cyclone (here, Typhoon Malakas of September 2010) eastward across the Pacific. A more general composite (averaged) study of many cases this phenomenon can be seen in a journal article by Torn and Hakim (2015).

Open the `LMT_1.5` IDV bundle. You will see a globe view as in Fig. 1.3 below, with 850-hPa heights displayed in yellow (restricted to only the western Pacific), and global 200-hPa flow depicted in several displays. Absolute vorticity is in red shading, while a wide red contour shows the (highly smoothed) $\zeta_{abs} = 10^{-4}$ s^{-1} isoline. The dashed red contour shows where planetary vorticity $f = 10^{-4}$ s^{-1}: that is, the ~40°N latitude circle. Wherever the solid curve is south of the dashed curve, there is generally positive or cyclonic ζ_{rel}, and the opposite for northward excursions of the contour. Rossby waves can be viewed as excursions of this contour, acting a bit like a plucked string on a musical instrument.

The bundle opens at 0000 UTC 25 September 2010, when TC Malakas (seen as a closed contour in the 850-hPa geopotential heights) is transitioning to midlatitudes, with its upper-level flow acting to "pluck the string." Rossby waves propagate eastward from this event, arguably affecting weather over North America.

a) Explore the displays. Check the boxes to display 250-hPa height (`global hgt`), and `Eddy height` (where *eddy* is defined as deviations from the zonal average around the whole planet along each latitude line). Step forward to 1200 UTC 25 September. **Capture images to show how the eddy height field relates to the north–south excursions of the solid red contour. Describe the relative locations of the northward-moving Tropical Cyclone Malakas and the upper-level ridge located to the north and east of it at this time.**

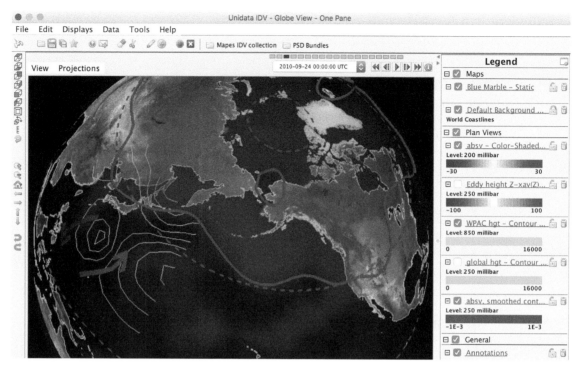

Figure 1.3. Screen capture of LMT_1.5 bundle.

b) Step the time ahead by a few more days to **identify a time with a large-amplitude downstream wave (trough–ridge couplet)**, and step forward and backward in time to gain a sense of the upper wave evolution. **Estimate the longitude of the ridge that forms immediately downstream of Malakas at 1200 UTC 25 September, and track this feature until 0000 UTC 28 September.** Use the readout of `Longitude` of the cursor at the bottom of the IDV window. **Estimate the eastward speed of this ridge, utilizing your knowledge of the spacing of longitude lines at the latitude in question. State any approximations or sources of possible error** in your calculation.

c) Notice that the amplitude of the ridge weakens by the end of the time period specified (0000 UTC 28 September). This indicates that energy is dispersing downstream at the Rossby wave group velocity, a topic discussed further in Chapter 2 of MSM (see its Fig. 2.20 for example). As discussed in section 1.5.3 of the MSM text, a useful and compact graphical means of representing Rossby wave activity is the longitude–time section or *Hovmöller diagram* (Fig. 1.4 below). **Utilize Fig. 1.4 to locate the ridge associated with Malakas (which you tracked in b.) at the 500-hPa level. Compute the *phase speed* of this ridge,** which is the slope of the orange streak on this diagram. **Compare this to the value you estimated in b.) Is it consistent? Discuss the sources of uncertainty in these calculations.**

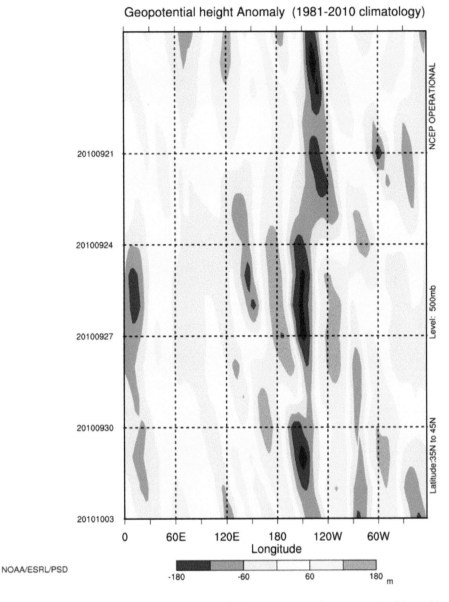

Geopotential height Anomaly (1981-2010 climatology)

NCEP OPERATIONAL

Level: 500mb

Latitude:35N to 45N

NOAA/ESRL/PSD

-180   -60   60   180 m

Figure 1.4. Hovmöller diagram (created with http://www.esrl.noaa.gov/psd/map/time_plot/) depicting 500-hPa geopotential height anomaly between 35 and 45°N from 17 September through 3 October 2010. Note: An anomaly is a deviation from the time-mean long-term climatology, while an eddy field in the IDV bundle is a deviation from the zonal (longitude) mean at an instant. These are not equal, but they both expose similar patterns.

d) Figure 1.4 indicates a series of alternating troughs and ridges that appear downstream of Malakas. **Sketch a line approximately connecting a set of maximum *amplitude* features in this "wave train" of troughs and ridges. Compute the *group velocity* implied by this wave train (or "packet") based on the slope of your line. Which is the larger eastward speed, the phase or group velocity?**

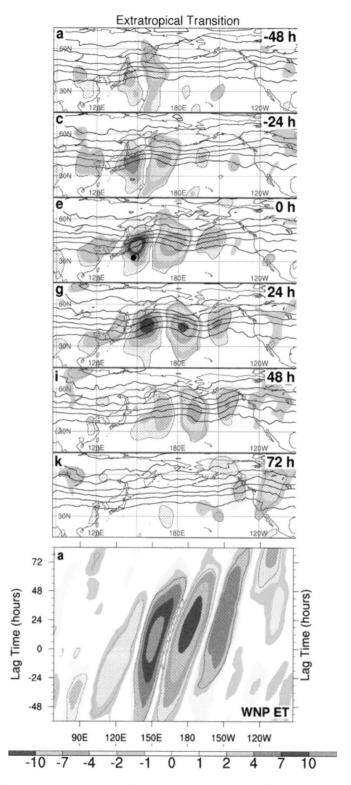

Figure 1.5. Composite map sequences and time–longitude sections of meridional wind anomalies averaged between 35° and 55°N for extratropical transitions of tropical cyclones in the western Pacific. Adapted from Figs. 2 and 3 of Torn and Hakim (2015). Note that time runs upward on the bottom panel, but downward in the map sequence shown and in Fig. 1.4 above.

e) In the IDV, rotate the globe to see remote regions (over North America) more clearly. **How far downstream in the midlatitude jet stream do you think you can detect waves emitted by Malakas? Make the case** using an image sequence you capture, annotating the features you feel are linked. **Compare these results with the Hovmöller (time–longitude) plot in Fig. 1.4, and discuss the comparison.** Examine Fig. 1.5 (which is taken from Figs. 2 and 3 of Torn and Hakim), an average of many cases. The Web link to this paper is included at the end of the chapter. **Relate your findings to that composite**, which will help ensure you are not over-interpreting the data in this one case, since Malakas was not the only weather event or factor in the hemisphere causing troughs and ridges to develop over North America. **In the IDV bundle, can you identify a source for the following Pacific "wave train" (lower part of Fig. 1.4), after the Malakas-triggered events?**

f) **Optional**: Examine other historical cases in 1979–2016 with the same displays, by using the *LMT_1.5_MERRA_1979-recent* IDV bundle, adjusting the date as explained in exercise 1.3 f) above, and the Web link in the caption of Fig. 1.4.

Reference

Torn, R. D., and G. J. Hakim, 2015: Comparison of wave packets associated with extra-tropical transition and winter cyclones. *Mon. Wea. Rev.*, **143**, 1782–1803, doi: http://dx.doi.org/10.1175/MWR-D-14-00006.1.

2

QUASIGEOSTROPHIC THEORY

This chapter includes the following exercises:

2.1. Conceptual View of QG Omega: Forcing and Response
2.2. The Traditional QG Omega Equation: Forcing for Ascent during a Winter Storm
2.3. Q-vector In-Class Exercise
2.4. Q-vectors Displayed Using Case-Study Data
2.5. Real-Time Forecast Discussion: Utilize QG Thinking
2.6. QG Height Tendency Equation Exercise
2.7. Potential Vorticity Exercise

Each exercise in this manual uses these typefaces for clarity:

Normal typeface is used for background information, technical instructions, motivating questions, and learning objectives. **Bold indicates assigned actions and questions that students are expected to respond to in their report.** A `constant width` typeface is used to indicate text that can be found exactly on the IDV software (usually on the `Dashboard` or `Legend` areas).

The word **Optional:** is used to set off suggestions for further explorations.

Around the time when computers were first becoming viable for numerical weather prediction, leaders in atmospheric dynamics such as Jule Charney recognized the need to simplify the predictive equations in order to save computational expense, while retaining the dynamical essence of important weather systems (e.g., Charney 1948, 1950). While the barotropic vorticity equation for a single-layer flow (such as at 500 hPa where the horizontal divergence term in the vorticity equation is often small) offered some prognostic skill, midlatitude weather systems of interest most often arise in highly *baroclinic* atmospheric states, with vertical structure (such as wind shear in thermal wind balance with temperature gradients as in Chapter 1) that requires a multilayered (three dimensional) viewpoint.

Although actual synoptic-scale wind patterns are near a state of geostrophic balance, we would not want to base a model on equations that specified *exactly* geostrophic flow. Recalling that geostrophic balance throws away the d/dt term, and that geostrophic flow has (almost) no divergence, there would be no clouds and precipitation, and no changes with time to try and forecast, in a truly geostrophic atmosphere! As a compromise, the "quasi-geostrophic" (QG) equations assume that the ageostrophic component of the wind is so weak that it does not importantly advect air properties from place to place. Still, the ageostrophic wind is important: for supplying mass horizontally to and from areas of vertical motion, and for its contribution to making the (mostly geostrophic) winds change with time [Eq. (2.14) in the textbook].

The QG system solves for *one part of* the ageostrophic wind (and the omega field, related by mass continuity), the part that does one job. That job is to maintain thermal wind balance through time, in the face of advective, diabatic, and frictional tendencies that tend to disrupt that balance. The ageostrophic wind has some other components too: for instance, one component is aligned against the geostrophic wind in curved flows that have sub-geostrophic speeds (in gradient or cyclostrophic balance, as in Fig. 2.2 in the textbook). Another component is the horizontal branch of the flow in unbalanced internal gravity waves. Still, the "maintainer of balance across time" component of ageostrophic flow that QG theory solves for is a major weather-maker of midlatitude storms and fronts, both directly (through the condensation of water in its updraft regions), and indirectly (through its role in spinning up the vorticity in cyclones).

Most importantly, the QG system of equations also offers insight into atmospheric dynamics. For weather forecasting, we can form equations for the QG vertical motion (the omega equation) or for the QG height tendency (a relationship that is even clearer in terms of potential vorticity, revisited in Chapter 4).

Chapter 2 in MSM presents quasigeostrophic theory. Here, we provide an accompanying set of exercises emphasizing the application of QG material in the interpretation and prediction of midlatitude weather systems.

2.1. Conceptual View of QG Omega: Forcing and Response

The QG omega equation is shown below. It states that *forcing for vertical motion* is related to the *differential vorticity advection* and *the Laplacian of the temperature advection*. For both advections, the approximation is made that the geostrophic wind is a good enough approximation for the advecting velocity. Recall that the change in geopotential Φ with pressure in the second term (the thickness dZ per unit mass layer dp) is a measure of layer-averaged temperature. Recall also that the Laplacian (∇^2) of the geopotential in the first term is proportional to geostrophic relative vorticity (with some constants). These ideas may help you to recognize the terms in the omega equation.

$$\left(\nabla^2 + \frac{f_0^2}{\sigma}\frac{\partial^2}{\partial p^2}\right)\omega = \frac{f_0^2}{\sigma}\frac{\partial}{\partial p}\left[\vec{V}_g \cdot \nabla\left(\frac{1}{f_0}\nabla^2\Phi + f\right)\right] + \frac{1}{\sigma}\nabla^2\left[\vec{V}_g \cdot \nabla\left(-\frac{\partial \Phi}{\partial p}\right)\right] \qquad (2.29)$$

Our objective in this lesson is to understand this diagnostic relationship. In other words, *why* does cyclonic vorticity advection increasing with height, or a horizontal maximum of warm advection, represent forcing for ascent?

This example is inspired by the Durran and Snellman (1987) paper. Consider an idealized jet streak situation. Suppose the 1000-hPa geopotential height surface $Z_{1000} = 0$ everywhere (i.e., the 1000-hPa surface is flat, with no geostrophic wind at that level). On the following diagram, the 500-hPa height contours are shown as bold solid lines, and the 500-hPa isotachs are shown as dashed lines. The cross section B-C is marked for later reference.

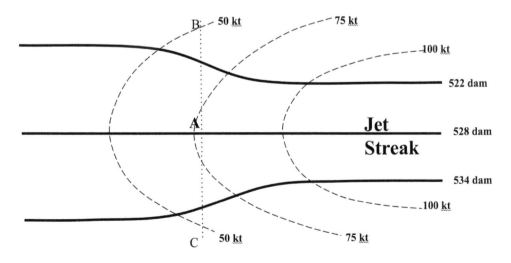

a) From the QG momentum equation [Eq. (2.9) in MSM], the local tendency of the geostrophic wind speed is in part *due to geostrophic advection* [evident if one expands the total derivative on the left side of Eq. (2.9)]. Similarly, we could expand the total derivative on the left side of the QG zonal momentum equation [Eq. (2.14)]. **What is the sign of the term at point A?**

$$-u_g \frac{\partial u_g}{\partial x}$$

How would this term tend to change the strength of the vertical wind shear with time (increase or decrease) at point A?

b) Next, consider the *geostrophic advection of temperature* in the lower troposphere. **Is there warm, cold, or neutral advection (+, –, or 0)? Why?**

Explain: **What information do we have about *temperature* on this diagram? What is the thermal wind in the 500–1000-hPa layer? If thickness and height contours are parallel, what does that imply about temperature (thickness) advection?**

c) Recall that thermal wind balance relates the vertical shear of the geostrophic wind to the horizontal gradient of temperature, at every instant in time. Here,

$$u_{g\,U} - u_{g\,L} = -\frac{C}{f} \frac{\partial \overline{T}_v}{\partial y}$$

This is Eq. (1.44) in MSM. Here, the subscripts "U" and "L" correspond to "upper" and "lower" levels, and "C" is a constant for a given pressure layer. **Would thermal wind balance be sustained in locations such as point A as the advective tendencies from a) and b) act together? Why or why not?**

d) If the answer to c) is "no," **explain the sense of the imbalance** that would develop at point A. In other words, would the vertical shear of the westerly flow become too weak for the north–south temperature gradient, or vice versa?

e) **Based on your answer to d), what would need to happen in order to bring the atmosphere back toward thermal wind balance in the vicinity of point A?**

The magnitude of the temperature gradient would need to _____

AND/OR

The magnitude of vertical wind shear would need to _____

f) What sort of vertical circulation (in this case, in the y–z cross-sectional plane) would be required to bring about the needed changes (that is, to restore or maintain thermal wind balance)? Remember, a "circulation" needs to satisfy mass continuity. Consider the vertical cross section below corresponding to the B-C cross section above. The perspective is from the west (i.e., you are standing with your back to the wind, looking eastward).

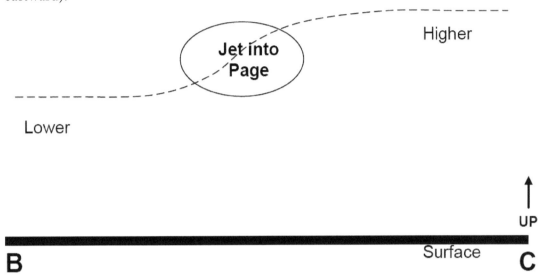

Sketch the ageostrophic circulation (using vertical and horizontal arrows) that would be needed to bring the atmosphere back towards thermal wind balance after advection has acted for a moment to disrupt that balance. Hint: One useful approach is to think first of how the horizontal wind would need to change with time. Recall from Eq. (2.14) in the text that $du_g/dt_g = fv_{ag}$ (the Coriolis force acting on the ageostrophic wind), so you can deduce what v_{ag} is needed. From v_{ag}, mass continuity then implies a vertical motion to close the circulation.

g) **Does the adiabatic cooling or warming effect of your deduced vertical motion help or hinder the action of fv_{ag} on the momentum field, in the overall two-part effort [as mentioned in e) above] to adjust the situation back to thermal wind balance?**

h) On the jet streak diagram below, **sketch a few vorticity contours, and label regions of cyclonic (use a C) and anticyclonic (use an A) shear vorticity.** Consider the wind orientation, based on the height contours. Combining the wind and vorticity information, **indicate areas of positive and negative vorticity advection.** Given that wind speed is increasing with height in this example, and therefore so is the strength of the advection, **where would you expect rising and sinking motion at the 700-hPa level (based on the first term in the traditional form of the QG omega equation)? Label this area on the diagram.**

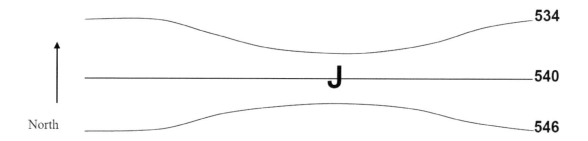

i) **Are the areas of ascent and descent you deduced above in a–f for the jet entrance, arising from the *thermal advection* term of Eq. (2.29), consistent with the ascent and descent you deduced in h) from the *vorticity advection* term? In other words, do the two forcing terms work in the same sense, or the opposite sense,** for this case involving self-advection of a geostrophically balanced jet streak?

j) **Optional:** Based on your thinking from the above, **can you imagine a geostrophically balanced flow situation in which the two terms in Eq. (2.29) cancel each other, so that no net QG omega forcing is present?**

2.2. The Traditional QG Omega Equation: Forcing for Ascent during a Winter Storm

The objectives of this exercise are to examine QG forcing mechanisms during a winter storm event and relate these to patterns of vertical motion and cloud cover. For familiarity, we will utilize the same case event from the thermal wind exercise in section 1.3.

Open the LMT_2.2 bundle. You should see displays of `Sea Level Pressure`, `Geopotential height`, and `Wind barbs` at 500 hPa. A magenta square shows the location of a movable `Sounding` probe.

a) Thermal advection at 1200 UTC 18 December 2009

   i. Consider the cyclone in the northern Gulf of Mexico. Based on the angle between the sea-level isobars and 500-hPa height contours, **where do you expect the most pronounced temperature advection at this time? In which areas would the QG omega equation predict forcing for ascent, if we were to base our analysis only on this term? Capture and annotate images to illustrate your answers.** Recall that the equation includes the *Laplacian of* the temperature advection field, so it is not just the *value* of temperature advection itself that is critical, but rather *local maxima* (where the Laplacian is negative) *or minima* (where the Laplacian is positive).

ii. Activate the `QG Temp Adv` display. **Was your intuition correct? Discuss. Use the sounding probe to confirm wind veering vs. backing in the warm vs. cold advection areas, as in exercise 1.3.**

b) Differential vorticity advection

The vorticity advection forcing term is mainly contributed by *upper-level* (500 hPa) vorticity advection, since winds are weaker at low levels.

i. Turn off the `QG Temp Adv` display. Select the `Absolute Vorticity` field in the legend and notice its relationship to the troughs and ridges in geopotential height. **Based on the wind and vorticity fields, where do you expect the most prominent areas of cyclonic vorticity advection? Where are these areas in relation to the locations with significant warm advection?**

ii. Activate the `Geo Avor Advection` display (it is under the `QG diagnostics` category in the legend). **Are the regions of cyclonic vorticity advection consistent with what you expected? Capture images of the comparison between the two QG advection terms, and discuss them.** Since Eq. (2.29) states that the *Laplacian* of omega equals the forcing terms on the RHS, and inverting the Laplacian acts like a smoothing operation, let's smooth the forcing term. Click the blue `Geo Avor Advection` hyperlink in the legend and change the smoothing parameter to a factor of 5. Is this a better fit to the scales of true omega (ascent) suggested by the IR satellite imagery? To answer this, toggle the `IR satellite display` on and off to see the observed cloud tops field. **Capture images to illustrate your assessment. Discuss what aspects of the IR cloudiness display (convective clouds vs. cirrus decks) might be most related to vertical velocity below the 500-hPa level where the vorticity advection term was evaluated.**

c) Vertical profiles at 1200 UTC 18 December 2009

i. Consider the skew-*T* profile in the separate `Sounding_wind` window. Recall that we examined vertical profiles for this case in exercise 1.3a, for geographical reference. Place the skew-*T* marker in the `Map View` window over Louisiana. **Do you expect QG forcing for ascent or descent in this location at this time? Discuss and justify your answer in terms of the two QG forcing terms.**

ii. **Is the midtropospheric dewpoint depression (gap between temperature and dewpoint curves) consistent with the omega implied by the QG forcing terms in (i) over central Louisiana? Discuss.**

iii. Now, move the sounding location to southern Georgia. **Do you expect QG forcing for ascent or descent in this location at this time? Discuss and justify your answer in terms of the two QG forcing terms.**

iv. **Are the midtropospheric temperature and dewpoint profiles (or the dewpoint depression) consistent with the omega implied by the QG forcing terms over southern Georgia? Discuss.**

d) Analyzed omega

The *actual analyzed omega* can differ substantially from the QG omega whose Laplacian equals the forcing terms according to Eq. (2.29) shown at the beginning of section 2.1. Discrepancies can occur because of all the non-QG terms in the equations, including topographic upslope or downslope flow, convection, and other components of the ageostrophic flow field.

In the `Map View` window, toggle on the temperature advection (`QG Temp Adv`), vorticity advection (`Geo Avor Advection`), and the 700-hPa omega field (`Omega - Color-Shaded`) to explore the relationships among the three patterns. **How well does the analyzed region of ascent match the location of QG forcing for ascent diagnosed in a) and b)? Does the analyzed omega better match the vorticity advection term or the temperature advection term? Discuss.**

e) Infrared satellite imagery

Does the `IR satellite imagery` display agree qualitatively with the analyzed omega, QG-predicted omega, the individual QG omega forcing terms, and the sounding's dewpoint depression? Discuss.

f) **Optional: Examine the same set of displays for another weather situation and answer the questions above again.** To do this for a historical case in 1979–2015, use bundle `LMT_2.2_MERRA_1979-2015`. For real time data, use `LMT_2.2_RT`. In either case, to view another area of the globe, zoom out until your desired region is visible and then hold the `Shift` key while rubberbanding a latitude–longitude box with the left mouse button depressed. Then, to adjust the time(s) desired, click the 🄸 icon in the animation control area and operate the `Define Animation Times` tab in the popup, reading menu items carefully to avoid excessively large data requests. Unfortunately, the sounding probe will not automatically follow the other displays for IDV versions before 5.3. In that case, you have to zoom out after the data finish loading and move the probe to the area in which your data have been fetched. *If you find a* *case you like, you can save it as a .zidv bundle, and feel free to contact an author of this manual if you wish to share it more broadly.*

g) **Optional:** Explore the excellent widget illustrating all the quantities and terms in the traditional QG omega equation at http://www.meted.ucar.edu/bom/qgoe/qgoe_widget.htm. It is free, but each student will need to register. An instructor's email can be set for quiz score delivery.

We gratefully acknowledge Prof. Jim Steenburgh, University of Utah, for a masterly IDV bundle used in developing aspects of this lesson.

2.3. Q-Vector In-Class Exercise

The objectives of this short lesson are to allow students to compute **Q** vectors for a simple flow, and connect the convergence or divergence of these vectors to forcing terms in the traditional QG omega equation.

The QG omega equation is a tool for diagnosis of the processes that give rise to vertical air motions on the synoptic scale. The "traditional" form of the QG omega equation utilized in the previous exercise is not always well suited for operational weather forecasting. As Durran and Snellman (1987) demonstrated, cancellation can occur between the right-side terms in the traditional omega equation. When the two terms on the right-hand side of the traditional form of the QG omega equation oppose each other, forecasters using that equation as the basis for their interpretation are faced with the difficult challenge of deciding which term is larger! Instead, it is advantageous to combine the right-side terms, as shown by Hoskins and Pedder (1980), to obtain the "**Q** vector" form of the QG omega equation:

$$\left(\nabla^2 + \frac{f_0^2}{\sigma} \frac{\partial^2}{\partial p^2} \right) \omega = -2\nabla \cdot \vec{Q}, \tag{2.30}$$

where

$$\vec{Q} = -\frac{R}{\sigma p} \begin{bmatrix} \dfrac{\partial \vec{V}_g}{\partial x} \cdot \nabla \theta \\[2ex] \dfrac{\partial \vec{V}_g}{\partial y} \cdot \nabla \theta \end{bmatrix} = \begin{pmatrix} Q_i \\ Q_j \end{pmatrix} = -\frac{R}{\sigma p} \begin{bmatrix} \dfrac{\partial u_g}{\partial x} \dfrac{\partial \theta}{\partial x} + \dfrac{\partial v_g}{\partial x} \dfrac{\partial \theta}{\partial y} \\[2ex] \dfrac{\partial u_g}{\partial y} \dfrac{\partial \theta}{\partial x} + \dfrac{\partial v_g}{\partial y} \dfrac{\partial \theta}{\partial y} \end{bmatrix} \tag{2.31}$$

Recall that the QG vertical motion is a response to thermal wind balance disruption. It should therefore make sense that the "forcing" for vertical motion on the right side of the omega equation is related to gradients of wind and temperature, evident in the **Q** vector. In fact, the **Q**-vector form of the omega equation is derived by examining the difference between equations that expressed the two different components of the thermal wind relation. This difference reflects advective tendencies that would tend to disrupt balance, and it acts as a "forcing" for circulations that would act to push the atmosphere back towards a state of thermal wind balance.

In Eq. (2.31), the two components correspond to east–west- and north–south-oriented vector components. By evaluating the derivatives and products, we can determine the orientation of **Q** and ultimately locate regions of converging and diverging **Q**, which are clearly of meteorological interest from Eq. (2.30).

In order to illustrate **Q** vectors in a simplified idealized setting, consider the diagram below.

- Define the x axis to be parallel to isentropes, with cold values to the north ($\partial\theta/\partial y < 0$).
- Note that $\partial\theta/\partial x = 0$ in this idealized example (but not in general).
- Assume that v_g at some level varies sinusoidally as indicated in the diagram below.
- For simplicity, ignore the factor $-R/\sigma p$ in the **Q** expression [Eq. (2.31)]—but notice that it is negative.
- **Evaluate Q at each of the indicated points A–E, using "finite difference" techniques.**

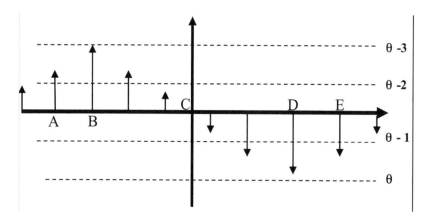

$$Q_i = -\frac{\partial u_g}{\partial x}\frac{\partial\theta}{\partial x} - \frac{\partial v_g}{\partial x}\frac{\partial\theta}{\partial y}$$

$$Q_j = -\frac{\partial u_g}{\partial y}\frac{\partial\theta}{\partial x} - \frac{\partial v_g}{\partial y}\frac{\partial\theta}{\partial y}$$

a) **Where would you find ascent? Is this where you would expect it based on the traditional form of the QG omega equation? Explain and discuss.**

b) **Optional: Sketch a different situation, for example involving *westerly* winds and isentropes at some angle,** a more realistic situation than that depicted above. It will help you to simplify the math if you still define your x and y axes so that $\partial\theta/\partial x = 0$, rather than using x as east and y as north. To do this, **sketch a realistic weather situation, then rotate the paper and add the axes** to think about the terms. **Evaluate the Q vectors, and their convergence.**

2.4. Q Vectors Displayed Using Case-Study Data

This exercise examines and interprets **Q** vectors and their divergence at the 700-hPa level, again focusing on the December 2009 case used in exercise 2.2. The objective is to compare and understand how the traditional QG forcing terms relate to the **Q**-vector forcing. Recall from Eq. (2.30) in MSM and from exercise 2.3 above that the right-hand side ("forcing") term in this version of the QG omega equation is proportional to $\nabla \cdot \mathbf{Q}$, the divergence of the **Q**-vector field.

In order to really solve for QG omega, the second derivatives on the left-hand side would have to be inverted via a numerical process such as successive over-relaxation. Qualitatively, such an anti-differentiation process greatly smooths the resulting QG omega field compared to the fine-grained $\nabla \cdot \mathbf{Q}$ field, as the opposite of differentiation (the Laplacian), which acts to enhance small-scale features in a field. As an approximate inverse Laplacian, simple smoothing is used on the $\nabla \cdot \mathbf{Q}$ displays in the bundles below. You can play with smoothing to see the noisy raw $\nabla \cdot \mathbf{Q}$ if you like.

a) Open the LMT_2.2 bundle that we used in section 2.2 above. Activate the displays for `Temperature contours`, `Hoskins Q-vector`, and `Wind barbs`. Confirm that all three displays are set to the 700-hPa level. **Examine the Q-vector field in the vicinity of the cyclone in the northern Gulf of Mexico. Capture an image that corresponds to the idealized diagram in problem 2.3 above. In what location are the vectors converging? Where are they diverging? Do these regions match your expectations from QG forcing terms in the traditional form of the omega equation from 2.2? Where are these areas relative to the upper-level trough, and surface cyclone?**

b) Deactivate the `Temperature contours` and `Wind barbs` displays, and activate the `Div(Q vector)` display, which is heavily smoothed as justified above.

   i. Compare the `Div(Q vector)` at 700 hPa to the `Omega` display at 700 hPa. **How well does the Q-vector convergence match up with areas of ascent (purple shading in both cases)? The Q vectors should point towards areas of ascent. Do they?**

   ii. Choose an area of notable **Q**-vector convergence, and **use the other QG diagnostics displays to attribute the forcing for ascent to either differential vorticity advection or temperature advection.**

   iii. Compare the displays of `Div(Q vector)`, `700-hPa omega`, and `IR satellite` imagery. **Discuss the degree of correspondence between these three plotted quantities. How well do the areas of Q-vector convergence** correspond to the convective and/or cirrus cloud decks evident in the satellite image? Notice the discrepancy in areas where convection is prominent. In Eq. (2.31) above, the definition of **Q** involves static stability σ. **What would happen to Q if condensational heating acts to make the *effective static stability* in some region much smaller than the dry static stability σ?**

    iv. Change the levels of `Hoskins Q-vector` and `Div(Q vector)` from 700 to 500 hPa and then 850 hPa (to do this, click on these displays' hyperlinks, which will bring up a window where you can choose from a list of isobaric levels). **Are the locations of Q-vector convergence sensitive to the choice of vertical level? What factors should go into determining which level to examine?**

    v. Zoom out to view a broader region and again compare the `Div(Q vector)` field to omega. **How good is the correspondence on broader spatial scales? Are there patterns to the discrepancy or agreement?**

c) **Identify an area with the "QG cancellation problem" (e.g., cold advection but cyclonic vorticity advection) based on the sea level pressure and 500-hPa height contours alone.** Use the `Drawing Control` in IDV (the pencil icon in the toolbar) to mark your location of cancellation. If you mismark, you can delete your marks (called glyphs) from the list under the `Shapes` tab in the `Drawing Control`. **Capture an image and indicate the area.**

d) Now, activate the `QG diagnostics` displays needed to check your answer from c) above. **Did you correctly select a location that experiences the QG cancellation problem? Which of the competing QG terms appears to dominate, based on Div(Q vector), the omega field, and IR imagery?**

e) Create a new display, an *isosurface of omega* enclosing the volumes of air with upward and downward motion.

To do this, follow this screen capture:

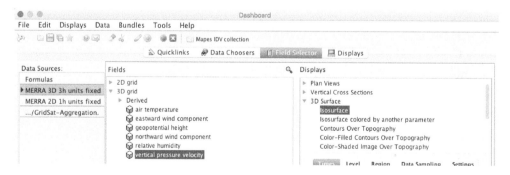

From the `Field Selector` on the `Dashboard`, highlight the `MERRA 3D 3h` data source under the `Data Sources` panel. Under the `Fields` panel, expand the `3D grid` item and choose `vertical pressure velocity` (the field name will appear as omega). In the `Displays` panel, expand the `3D Surface` item, and select `Isosurface`. Click on `Create Display`.

Now, in the `Legend` of the display window, click on your new `omega-Isosurface` and set the isosurface value to -0.5 Pa s^{-1} in the display control that pops up. The negative value corresponds to a region of *ascent*. Repeating the steps above, create a second isosurface of *descent*, with value $+0.25$ Pa s^{-1}. Adjust the color and transparency of your isosurfaces to taste.

Examine the isosurfaces in three dimensions in this cyclone. Is there a systematic tilt with latitude? Do you see any evidence in the IR imagery for the analyzed omega field being correct or incorrect? Discuss.

2.5. Real-Time Forecast Discussion: Utilize QG Thinking

QG reasoning can be applied to everyday weather situations, as part of a skill set for delivering effective, science-based weather briefings. This exercise works toward those goals by applying QG analysis and forecasting techniques to develop an *Area Forecast Discussion (AFD)* for the current and upcoming weather situation. Specifically, the objective of this lesson is for students to construct a forecast discussion based upon application of QG concepts to a current weather situation.

It is important to recognize that non-QG processes can exert a dominant influence on local weather. Convection and topographically forced processes are two examples. However, it is equally important to understand scale interactions, and knowing how the larger, synoptic-scale atmosphere is evolving is a prerequisite to understanding mesoscale or other non-QG processes. Likewise, recall from chapter 1 that Rossby wave packets are often hemispheric in scale, and when diagnosing QG processes related to a given trough or ridge it is helpful to step back and think about the planetary-scale context for the synoptic systems of interest. Ultimately, students of the atmospheric sciences must integrate information across many scales, and utilize understanding of planetary, synoptic, mesoscale, and perhaps microscale processes in research and forecasting applications. Here, our focus is on the synoptic-scale signals for which the QG equations convey useful understanding.

An important challenge that confronts atmospheric scientists is to boil a large volume of information down into a concise (and actionable) summary and forecast. As you will see in this exercise, vast amounts of information are available. However, not all of this information is equally important! Your task is to identify the important features, explain why they are evolving the way they are, and how they will affect the weather in a given location. Time does not permit equal treatment of all weather features!

PURPOSE OF AN AREA FORECAST DISCUSSION

The purpose of an AFD is to provide a well-reasoned discussion of the meteorological thinking behind a forecast (anticipated weather conditions) for the local county warning area (CWA) as used by U.S. National Weather Service (NWS) offices.

The AFD is a scientifically enhanced discussion with the goal of coordination between a local NWS office and adjacent NWS offices. The AFD is also a public product and is the primary means available to clearly convey the reasoning behind NWS forecasts to external users, including private meteorologists. The goal of the NWS forecaster is to write a concise, informative statement of forecast reasoning in order to inspire confidence in (or communicate uncertainty about) the forecast among the primary users.

CONTENT AND PHILOSOPHY

The text of an AFD is written in plain language and/or with the use of proper abbreviations. The text is as concise as possible, yet it covers the most significant characteristics of the forecast. There is no specific discussion length that is "right" for every weather situation. The AFD is a narrative description of the *scientific basis for forecast decisions*. **Your assignment is to write an Area Forecast Discussion for the assigned location.** NWS forecasters have limited time to write the AFD, and so this activity is well suited for an in-class exercise—*use your time efficiently!* **Consider the weather forecast for the assigned region for the 3-day period beginning at 1200 UTC today.**

Your AFD content should include answers to the following questions:

What is *currently happening* in the *assigned area of interest* and the surrounding region? Consider large-scale (e.g., North America or U.S.) perspectives, then zoom in on the assigned area of interest. **What is causing the current weather pattern? Limit your discussion to the main synoptic-scale weather systems. Base your discussion on QG reasoning to the extent appropriate.**

Which weather systems *will be* affecting the assigned area of interest *in the next 3 days*? What will be happening in the next 3 days in this area, and why? Why will the weather systems affect the area in the way that we expect? Discuss the evolution

of the synoptic pattern as it relates to the forecast. Again, use QG reasoning where appropriate. Be precise about the expected timing of events.

The following exercise can draw on information from any source, but here we provide three IDV bundles that will supply the needed information.

a) **[1 paragraph]** Load the `LMT_2.5_obsNA` bundle, which provides observational data (satellite, radar, surface observations, and model analyses) for North America. **Describe the current meteorological situation in the vicinity of the assigned location at 1200 UTC today**, with an emphasis on the weather systems that are affecting the assigned area, and which, based on extrapolation, you expect will be affecting the forecast area in the short term (next 24 h). **What is happening now, and what has happened recently? Where are the systems that *will be* affecting the local area *currently located*? Discuss, where possible, in terms of QG processes. For any prominent areas of cloud and precipitation, can QG processes explain their existence? Which, if either, of the traditional QG omega-equation forcing terms are responsible for these areas?** Activate the `Front Display`. **Are there any frontal systems near or approaching the forecast location?**

b) **[1 paragraph]** Load the `LMT_2.5_fcst` bundle, which shows a four-panel display of numerical model output. Loop the animation and get a feel for all the displays and their patterns. **What is shown in these four panels, and why? Examine the forecast, with an eye on QG diagnostics.** For example, 500-hPa height and vorticity, sea level pressure, and the QG diagnostics utilized in earlier lessons are all available. Using these, **provide a "QG discussion" of the *synoptic-scale pattern evolution*, processes, timing, and intensity of QG forcing for vertical motion. *Explain the expected impact of synoptic-scale weather systems on weather conditions at the assigned location*** for the forecast period and area. Focus on the 3-day forecast period, and be precise in specifying locations and times. You may wish to organize the discussion day by day. Utilize your knowledge of the QG omega equation to explain how weather conditions will evolve. Remember, this is about the scientific reasoning behind the forecast, not an actual forecast. Focus on processes.

c) **[1 paragraph]** No forecast is complete without a discussion of forecast uncertainty. Accordingly, load the `LMT_2.5_ens` bundle. Examine the GFS Ensemble Forecast System (GEFS) output to examine the degree of ensemble spread, which is one measure of confidence in a forecast. Consider any important troughs, ridges, or jets in the area of interest. The "spaghetti" contours show a given height contour for each of the different GEFS members, and the deviation between these contours is a measure of uncertainty in the forecast. The computed standard deviation is plotted with purple shading. Use this information to **discuss the level of forecast confidence you have for the key synoptic-scale weather systems you emphasized in your paragraph from b).** Note that there are many additional parameters available here, including temperature, sea level pressure, and precipitation. Utilize these as necessary in discussing forecast confidence.

d) **[1 paragraph] Are there any non-QG processes that should be taken into account for the forecast? If so, discuss what they are and how they could affect the forecast.**

2.6. QG Height Tendency Equation Exercise

Recall that the ageostrophic part of the wind is important for two reasons: (1) it contains all the horizontal divergence, the key to clouds and rain through the associated omega field, and (2) it generates the time evolution of the flow pattern, the key to forecasting. QG theory helps us estimate the part of the ageostrophic wind field that performs both of these jobs in synoptic-scale flow.

We saw above that the QG omega equation helps us interpret the instantaneous or *diagnostic* reasons for synoptic-scale vertical motions. In contrast, the QG height-tendency equation is *prognostic*, meaning that it can help us to understand why weather systems change with time. The objective of this lesson is to utilize real-data cases to apply the height tendency equation, with the goal of providing an explanation for the causes behind the time evolution of synoptic-scale systems. Review section 2.4 of MSM for derivations and interpretation of this equation.

The QG height-tendency equation [Eq. (2.34) in the book] is repeated here. Here χ is the time rate of change of geopotential height under QG assumptions. Remember that the Laplacian operator on the left side acts like a negative sign for simple waves, and that the process of inverting it to obtain the height tendency χ is like a smoothing operation, de-emphasizing small-scale features in the "forcing" terms on the right.

$$\left[\nabla^2 + \frac{\partial}{\partial p}\left(\frac{f_0^2}{\sigma}\frac{\partial}{\partial p}\right)\right]\chi = -f_0\vec{V}_g\cdot\nabla\left(\frac{1}{f_0}\nabla^2\Phi + f\right) - \frac{\partial}{\partial p}\left[-\frac{f_0^2}{\sigma}\vec{V}_g\cdot\nabla\left(-\frac{\partial\Phi}{\partial p}\right)\right] - \frac{f_0^2}{\sigma}\frac{\partial}{\partial p}\left(\frac{R}{C_p\,p}J\right)$$

The first right-hand term is the advection of vorticity, and tells us that ridges and troughs in the height field tend to drift downwind along with their associated anticyclonic and cyclonic vorticity. This is extremely unsurprising, since map discussions often just make the leap that "ridges and troughs are advected" by the jet stream. The second right-hand term involves the advection of temperature (noticing that $-\partial\Phi/\partial p = -g\,\partial Z/\partial p$ is thickness, proportional to temperature). Its triple negative sign is easiest to understand geometrically, by picturing wedges of thicker or thinner air being blown into the air column above or below the pressure surface (perhaps 500 hPa) whose height change is χ. The third term involves diabatic heating J, and again can be visualized as a thickening or thinning of the air column above or below the pressure surface in question.

a) Open the LMT_2.6 bundle in the IDV. Watch the loop of `Geopotential height` (500 hPa) and the most-positive values of `absvort` (absolute vorticity, magenta shading) to get oriented to the weather situation. **Describe the weather pattern in terms of the motion and evolution of the main three to four troughs and ridges at 500 hPa.** Activate the `Sea Level Pressure` display, and **relate the story of the main surface cyclones and anticyclones to this activity of the troughs and ridges at 500 hPa.**

b) **Loop and examine** the `Geopotential height` and `Time step difference` displays. [You may notice that the latter one, $Z500(t) - Z500(t-1)$, is not available at the first time step.] Step through the sequence a few times. **Where in relation to the troughs and ridges are the largest positive and negative 500-hPa height tendencies?** (You may notice, especially if you zoom out to see the tropics, that the time step difference field shows broad weak red and blue patches indicating westward-moving rises and falls of $Z500$. These are atmospheric tides driven by solar heating in the upper atmosphere. They should be ignored here).

c) Let's focus on the height falls associated with the eastern U.S. trough's development and motion. Notice that the trough *moves eastward, deepens,* and *sharpens* during the sequence. Can you ascribe these time changes to individual terms in the QG height tendency equation?

   i. Set the time to 0000 UTC 12 September, and activate the `GeoVortAdv` contour display. Recalling that *positive* vorticity advection is associated with *negative* height tendency, **show that height falls east of the trough axis correspond to the advection of vorticity features. To do this, capture images of a time when the `Time step difference` pattern shows the signature of eastward motion and agrees well with the geostrophic vorticity advection pattern.**

   ii. To see if thermal advection also contributes to the trough's motion and/or amplification, activate the `Sea Level Pressure` and `Geopotential height` contours only. **What is the sign of thermal advection in the surface to 500-hPa layer (inferred from veering or backing geostrophic wind, evident from isobars and height contours) in the trough? How does this thickness advection below the 500-hPa level contribute to the height falls in the base of the trough?** Consider both 0600 UTC 12 September and 0600 UTC 13 September. **Capture images, annotate, and discuss the role of this second term in the height tendency equation.**

iii. Activate the *3D Surface* (isosurface) plots of lower-tropospheric temperature advection (`pos., neg. T adv`). You may want to adjust the isosurface values. **Is there a correspondence between areas of lower-tropospheric cold (warm) advection and the region of negative (positive) 500-hPa height tendency associated with the trough in question?** Your answer should be generally consistent with the question above. **Capture images and discuss what you observe for both of the key times (0600 UTC 12 September and 0600 UTC 13 September).**

iv. Deactivate the `pos., neg. T adv isosurface` displays, and activate the isotach (`wind speed`) plot, along with the `GeoVortAdv` (geostrophic vorticity advection) contours. Late in the sequence, the trough sharpens, and a cutoff appears in *Z500*, causing an enhanced height gradient (and thus a strong geostrophic wind speed) on its east side. **Capture images of the sharpening process. Which physical process—vorticity or thermal advection—is more responsible for this sharpening?**

v. **Summarize** the evolution of the eastern U.S. trough within the context of the QG height tendency equation, with specific emphasis on the two key times (0600 UTC 12 September and 0600 UTC 13 September). **Does the QG height tendency equation appear to adequately describe this case example, specifically for these particular times? In other words, do the interpretations provided by this equation fit adequately (qualitatively) with the observed height changes?**

2.7. Potential Vorticity Exercise

The purpose of this lesson is to provide experience in utilizing potential vorticity (PV) to diagnose atmospheric structure and processes. You will interpret PV in relation to other atmospheric variables. Recall that PV is altered by vertical heating gradients. Positive PV values are generated either at the base of a heating process (at low levels, below latent heat release, for example in tropical cyclones) or at the top of a cooling process (like at the tropopause near the winter pole, atop the radiatively cooled troposphere, creating the planetary-scale polar vortex whose "tentacles" or filaments of PV are the essence of mid-latitude synoptic weather variability).

a) Load the LMT_2.7 bundle, a GFS forecast sequence. *This is a fairly data-intensive bundle, so give it some time to finish loading.* For the fields valid 1200 UTC 16

September 2014, **detect center locations and strengths of the following three intense cyclones: 1) over the North Atlantic, 2) near Baja California, and 3) beyond the western Aleutian Islands, near the edge of the data domain. Enter them in the table below.** To get the pressure values, utilize the Data Probe feature of IDV, following this screen capture:

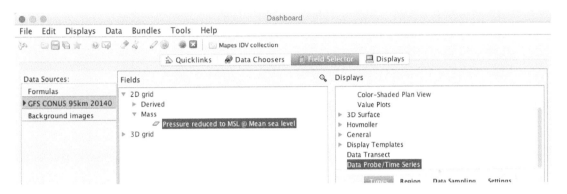

Verbal instructions: In the `Dashboard`, `Field Selector` tab, select the `GFS CONUS` dataset, `2D grid > Mass > Pressure reduced to MSL`. Under `Displays`, scroll down and select `Data Probe/Time Series`. Click on `Create Display`. **Move the probe, seen on the Map View window as a small red rectangle, to see time series at your cyclone center locations. An alternate method to retrieve point data is to middle-click with the cursor over the location of interest.**

Time/date of map: 1200 UTC 16 September 2014

Storm	Latitude of center	Longitude of center	Approximate GFS central pressure	Location description	Classification

b) Based on the sea level pressure, **what physical process(es) do you think led to the weather systems you identified in a) above? Are these tropical low pressure centers, extratropical cyclones, or can you even tell? If a recurring tropical cyclone moves to higher latitudes, how can you determine if it is of tropical origin or not?**

c) Lower tropospheric PV can be an indicator of past latent heating. In the `Map View` window under `Legend`, activate the `pvor - Isosurface` display, which is displayed only in the lower troposphere as a red isosurface. Also, select the `truewindvectors` (850 hPa) display. Use the mouse to tilt and rotate the display view as needed to gain a perspective on these weather systems. At this point, just use 1200 UTC 16 September.

    i. **What is the relation between the "PV towers" and the low pressure centers you identified in a)? Describe what you see**.

    ii. Toggle the display for `PV surface colored by height`, which shows the PV = 2 PVU surface that separates the generally low-PV troposphere from the generally high-PV stratosphere. **How can this information help you to distinguish tropical from extratropical cyclones? Discuss the linkages between the PV, physical processes, and weather system structure.**

d) In light of the PV, wind, and sea level pressure displays, **classify the weather systems at 1200 UTC 16 September that you identified in a) as "tropical," "extra-tropical," or "hybrid" in the last column of the table above**. Remember the physical processes that can lead to large PV in the lower troposphere, and that a weather system needn't be tropical in order to be accompanied by heavy precipitation.

e) Activate the `theta - Isosurface colored by height` display (the 315-K potential temperature surface, colored by its altitude). **Which of the weather systems you identified in a) are "warm core"? Which are "cold core"? Is this consistent with your classification?**

f) Turn off the rainbow-colored isosurfaces, and examine just the red `pvor - Isosurface` in the lower troposphere while stepping through time. Adjust the isosurface value to PV = 1 to maintain its identity (click on the display's hyperlink in the `Legend` to manipulate the value). Try to trace the remnants of Hurricane Odile (the one you marked near Baja) using this lower-tropospheric PV surface. **Where does this system go over time? Indicate the expected (forecast) location (in latitude/ longitude and geographical area) of this system at 0000 UTC 19 September and 0000 UTC 21 September.**

0000 UTC 19 Sep:

0000 UTC 21 Sep:

g) Deactivate all displays except for maps, then activate the `pvor - Color-Filled Vertical Cross Section` of PV, the `Geopotential height display` at 500 hPa, and `PV surface colored by height`. **Show the relationship between places where the PV = 2 surface is lowered and features of the Z500 field. Using the cross section, in an oblique view like the screen capture below, show that higher values of PV fill the atmosphere above these lowerings of the PV = 2 surface, and that the tropical PV towers built by latent heating are unconnected to the upper-level (stratospheric) reservoir of large PV.**

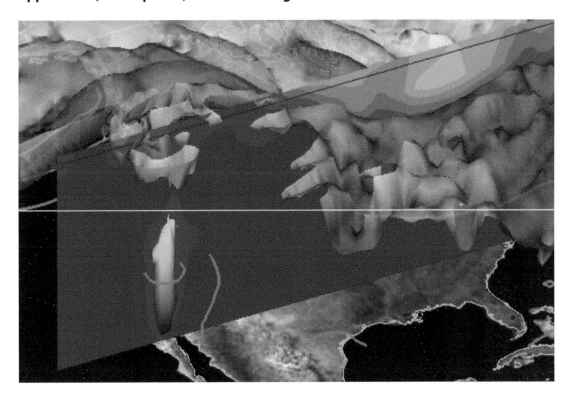

h) **Suppose that a more junior atmospheric science student were to approach you, and ask "What is this PV I keep hearing about, and how can I use it to learn something useful about weather systems?" Provide a short paragraph answering this question, in light of the exercise above.**

References

Charney, J. G., 1948: On the scale of atmospheric motions. *Geofys. Publ.*, **17,** 3–17.

Charney, J. G., 1950: Dynamic forecasting by numerical process. *Compendium of Meteorology*, T. F. Malone, Ed., Amer. Meteor. Soc., 470–482.

Durran, D. R., and L. W. Snellman, 1987: The diagnosis of synoptic-scale vertical motion in an operational environment. *Wea. Forecasting*, **2**, 17–31, doi:10.1175/1520-0434(1987)002<0017:TDOSSV>2.0.CO;2.

Hoskins, B. J., and M. A. Pedder, 1980: The diagnosis of middle latitude synoptic development. *Quart. J. Roy. Meteor. Soc.*, **106**, 707–719, doi:10.1002/qj.49710645004.

3

ISENTROPIC ANALYSIS

Chapter 3 presents isentropic analysis, an alternate method for the diagnosis of synoptic-scale vertical motion that is largely consistent with, but complementary to chapter 2's quasi-geostrophic (QG) approach to understanding vertical motion. Thinking about isentropic surfaces is also a useful viewpoint for analysis and interpretation of the atmosphere more generally. There is also a strong connection between isentropic concepts and the potential vorticity (PV) framework presented in chapter 4.

This chapter includes the following exercises:

3.1. Constructing an Isentropic Map
3.2. Isentropes and Trajectories in a Winter Storm
3.3. Comparison of QG and Isentropic Views on Vertical Velocity
3.4. Isentropic Techniques for Analysis and Prediction of Current Weather

Each exercise in this manual uses these typefaces for clarity:

Normal typeface is used for background information, technical instructions, motivating questions, and learning objectives. **Bold indicates assigned actions and questions that students are expected to respond to in their report.** A `constant width` typeface is used to indicate text that can be found exactly on the IDV software (usually on the `Dashboard` or `Legend` areas).

The word **Optional:** is used to set off suggestions for further explorations.

3.1. Constructing an Isentropic Map

The objectives and learning outcomes for this lesson are for students to 1) evaluate the information needed to construct isentropic charts from rawinsonde data, 2) manually construct an isentropic analysis from observational data, and 3) relate features on an isentropic chart to basic weather systems and their structure.

In order to appreciate a map of pressure (or any other quantity) on an isentropic surface, it is helpful to first gain a sense of how these maps can be constructed from observational data. To analyze pressure on an isentropic surface, we begin by using a skew-T diagram to determine the pressure of a given isentropic surface at a fixed point. Figure 3.1 shows a skew-T diagram from Greensboro, NC (KGSO) from 0000 UTC 3 January 2002.

a) i. **Using Fig. 3.1, determine the approximate pressure value corresponding to the 280-K isentropic level:** _____ hPa.

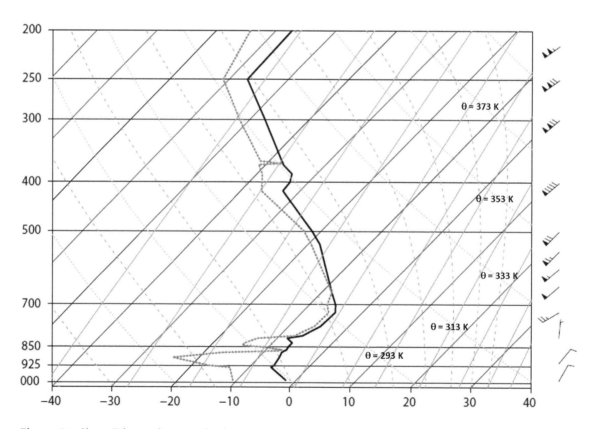

Figure 3.1. Skew-T, log-p diagram displaying rawinsonde data from Greensboro, NC (KGSO) valid 0000 UTC 3 January 2002. Potential temperature labels are shown along various isentropes (green dotted lines sloping upward toward the left).

ii. **Now, do the same for the 292-K level**: _____ hPa.

iii. **Next, locate the 330-K level**: _____ hPa

Suppose that we wished to generate a plot that shows the "topography" of a given isentropic surface, through contours of either pressure or height on a horizontal map. From a set of soundings, we would then plot the respective pressure value for a given isentropic surface (just like the ones you estimated above) corresponding to each sounding location. Figure 3.2 shows the values of pressure and horizontal wind corresponding to the 292-K isentropic surface from 0000 UTC 3 January 2002.

b) **Mark the location** of Greensboro, North Carolina (KGSO) on Fig. 3.2. Compare the value of pressure shown on the map at GSO for this 292-K surface to the value you estimated in a) ii. **Are the values close? If not, why not?**

c) **Draw solid isobars with an interval of 100 hPa on Fig. 3.2.**

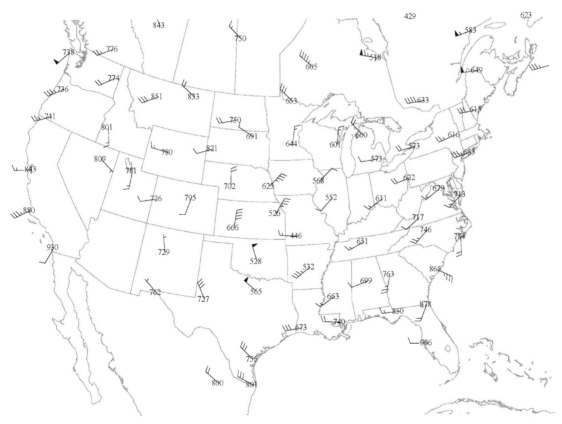

Figure 3.2. Values of pressure (numerical values, hPa) and wind (barbs) on the 292-K isentropic surface from standard rawinsonde sites at 0000 UTC 3 January 2002.

d) Inspect your analysis. **Where is this isentropic surface at higher or lower elevation? How does that relate to locations of warm and cold air?** Consider the sense of the wind and circulation, noting any centers of cyclonic or anticyclonic rotation. **What kind of weather system (warm or cool core, cyclone or anticyclone) is located over the center of the United States? Discuss and explain your reasoning.**

e) **Why are there no observations showing in a region of the southwestern United States near Arizona? Are there no rawinsonde stations there? Explain.**

3.2. Isentropes and Trajectories in a Winter Storm

Isentropic surfaces (also termed *theta* or θ surfaces) can be defined everywhere, at any instant, as in problem 3.1, except where the surface temperature exceeds the θ value of a given surface (that is, the surface is "underground" and fictitious).

Two main properties of potential temperature (θ) make it useful to think about θ surfaces. First, θ *always increases with height in a statically stable atmosphere.** As a consequence, θ surfaces are stacked vertically, and can be used as a vertical coordinate. Second, θ is conserved for adiabatic motion. Adiabatic flow is often an accurate approximation for anticyclones and other areas of the atmosphere without strong upward air motion, and it is sometimes a reasonable approximation for winter storms where condensation, radiation, and other diabatic heating and cooling effects are much weaker than temperature advection by the storm's winds. In that case, *air parcels tend to stay on the same theta surface*, so that vertical motion can be qualitatively estimated at an instant by assessing whether the horizontal winds are carrying air parcels up or down the local slope of their θ surface. The weakness of this view is that the topography of θ surfaces is itself always moving and changing. Still, the upward slope of θ surfaces toward the pole tends to prevail in most circumstances, so that poleward flow tends to ascend. Also, storm-related topographic features of a θ surface can sometimes be viewed as having a constant shape, translating with a constant speed (the "frozen wave" approximation; see MSM section 3.3). In this view, the upslope or downslope component of wind *relative to these long-lived but perhaps moving slope features on a θ surface* can still be interpreted as indicative of vertical motion, giving us a way of predicting or explaining cloudy versus clear weather conditions.

---

* In well-mixed layers of neutral static stability, potential temperature is constant with height. If it decreases with height, that layer would immediately undergo vigorous overturning or mixing. However these exceptions are rarely found in synoptic-scale weather systems, in midlatitude locations, above the planetary boundary layer.

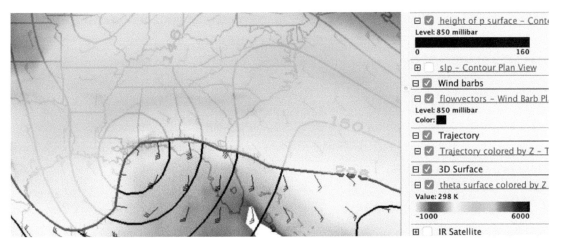

Legend items (right panel):
- height of p surface – Conto
  - Level: 850 millibar
  - 0 ———— 160
- slp – Contour Plan View
- Wind barbs
- flowvectors – Wind Barb Pl
  - Level: 850 millibar
  - Color: ■
- Trajectory
- Trajectory colored by Z – T
- 3D Surface
- theta surface colored by Z
  - Value: 298 K
  - -1000 ———— 6000
- IR Satellite

Figure 3.3. The LMT_3.2 bundle's initial display at 1200 UTC 18 Dec 2009. The 298-K `theta surface` is shown (semi-transparent, with color indicating its altitude in meters), along with the `850 hPa height of p surface` contours (in decameters) and winds (`flowvectors`) in black. Faint red plus symbols (+) indicate the starting points for 3D `trajectories`, initially located along the intersection of the 298-K and 850-hPa surfaces (red contour).

The objectives of this lesson are to explore air parcel trajectories in the vicinity of a winter storm, and evaluate their consistency with the idealized "conveyor belt" view of air gliding upward and downward along the sloping isentropic surfaces around a cool-core mid-latitude cyclone. Learning outcomes include 1) comparing 3D air parcel trajectories to the adiabatic approximation that air stays on isentropic surfaces and 2) interpreting vertical motion and trajectories within the context of the conveyor belt model.

Launch the IDV and open the bundle named LMT_3.2 (Fig. 3.3). You may recognize the case (18–21 December 2009) from the LMT_2.2 bundle from chapter 2. Notice the small red crosses along the line where the 850-hPa surface meets the 298-K surface (the heavy red contour). These points at 850 hPa are the starting points for 3D air parcel trajectories.

West of the low center, the horizontal winds are blowing toward the south or southeast, a direction that moves air *down* the slope of the isentropic surface. East of the center, the winds are blowing northward, into (or *up*) the slope of the isentropic surface. By advancing the animation's time control, you will see how well the altitude of the 3D trajectories (expressed on the same color scale as the height of the 298-K surface) supports the idea that air tends to conserve its $\theta = 298$ K as it flows along the sloping isentropic surface.

a) **Where would you expect upward air motion at the 850-hPa level based on this isentropic image at the initial time?** To check your expectation, **switch on the display for `omega`** at the 850-hPa level (you may need to turn off the `theta surface` to improve visibility). **Do areas of ascent match expectations? Explain.**

b) Zoom out and step through the time sequence. **How well does 850-hPa omega along the thick red 850-hPa–298-K intersection line agree with the isentropic reasoning? Are there places and times where the omega is opposite in sign to that from isentropic reasoning?**

c) Set the time to 0000 UTC 20 December 2009. Rotate the display to see a 3D view of the trajectories and isentropic surface, like the view from the south shown below (this icon in the left toolbar will jump you to the south view: ▣)

    i. For the eastern portion of the trajectories, note that they rise above the 298-K isentropic surface. **What physical processes could explain these non-isentropic air parcel excursions?**

    ii. **How did some of the deep blue trajectories get so far above the isentropic surface?** Jump back to the top view (Home icon ⌂) and find the time and place where these steep ascents occurred. Does the `IR satellite` display give any indications about the weather in that area?*

d) **Optional:** Let's look at vertical motions at another altitude.

To do that, change the level of the `height of p surface`, `flowvectors`, and `Omega` displays to 600 hPa, by clicking each of these three links in the `Legend` and adjusting the Levels menu in the corresponding Display Control on the `Dashboard`. Then change the 3D surface theta display to an isosurface value of 310 K, again by clicking its `Legend` link and adjusting its Display Control. **Repeat b) above. How well does "flow up and down the isentrope" reasoning explain the omega field at 600 hPa in this case? Discuss.**

For your new pressure and theta levels, **launch new trajectories** along the intersection of the new *p* and theta surfaces. To do this, hit the Home ⌂ icon to center the display, and reset the display time to the initial time. Edit the `298K on 850hPa` fields to instead be `310K on 600hPa`. Turn on the `theta surface` display so the intersection line is clear, and turn on the trajectory display so that you will see your results. To create new trajectories, click the `Trajectory` display's blue link in the `Legend` to bring up its `Controls` in the `Dashboard` window. First, use the `Levels` menu to select the new starting altitude (600 hPa). For `Trajectory`

---

* But don't over-interpret small details: the omega field may not be well analyzed—that is, the omega data shown (from MERRA reanalysis) may not be accurate for all the subsynoptic-scale features of IR imagery.

`Initial Area`, choose `Points`. Now go to the `Map View` window and use the left mouse button to select a set of points along the intersection of your omega field at 600 hPa and the theta surface. *Your mouse clicks won't be visible, so avoid over-clicking.* About 10–20 points should suffice.

When you have enough points, zoom out (since trajectories are not visualized outside your initial view region), and click `Create Trajectory`. Now as you advance the time counter, your trajectories should be visible. Some may be hidden by other displays, so toggle off other display layers to see more clearly. To change trajectory colors, click the `Trajectory` display in the `Legend` to bring up the display control in the `Dashboard`. Click the `Color Table` button and `Change Range` to [−1000, 6000] to make the trajectory colors match the coloration of the height of the theta surface. Other ways to aid visualization include jumping to a side view with ▱ or ▱, and rotating the display by dragging with the right mouse button pressed.

Do you find air parcels that go in surprising directions? Do more parcels go far above the isentrope than far below it? Can you explain these excursions, in terms of the properties of diabatic processes?

e) Let's look at more trajectories to see if we can reproduce the schematic of the *warm conveyor belt* pictured below in Fig. 3.4. In this view, upper-level outflow to the east of the storm comes partly from the warm sector where it ascends along the warm conveyor belt, and partly from the upper-level flow to the west. Turn off all the displays other than basic meteorology (heights, sea level pressure, maybe winds, and perhaps 850-hPa specific humidity to see low-level moisture tongues), and let's try to find the conveyor belt using only those displays for orientation.

   i. Create a swarm of forward trajectories from low levels at the initial time, using a `Rectangle` for the `Trajectory Initial Area` in the `Trajectory` display controls. **Did you find the warm conveyor belt? If so, capture an image illustrating this.**

   ii. Create a swarm of `Backward trajectory` displays, from a region where the `IR Satellite` shows a high (cold) cloud mass, again using a `Rectangle` for the `Trajectory Initial Area` and setting the `Levels` to 200 hPa in the `Trajectories` display control. **Show the results. Are there both horizontal and vertical airstreams contributing to the cloud mass as suggested in this cartoon? Capture an image illustrating this.**

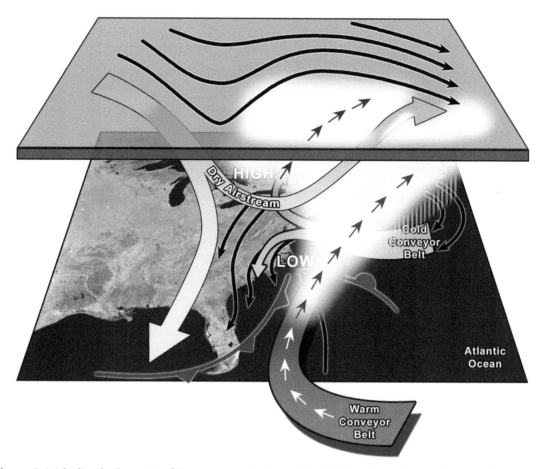

Figure 3.4. Idealized schematic of the conveyor-belt model of airflow through a northeast U.S. snowstorm. Adapted from Kocin and Uccellini (1990, their Fig. 26), based on Carlson's (1980) Figs. 9 and 10. Lines with arrows represent streamlines, colored airstreams indicated with annotations. White shaded area depicts cloud shield.

f) **Can you identify an air stream that corresponds to the "cold conveyor belt"?** You may need to begin at a later point in time. Again, capture an image if you can identify this air stream.

3.3. Comparison of QG and Isentropic Views on Vertical Velocity

The objectives of this exercise are 1) to relate the "traditional" QG vertical velocity techniques and interpretations from chapter 2 to the isentropic framework, and 2) to examine a challenging forecast situation using the isentropic analysis framework.

a) **Load the LMT_2.2 bundle**. It includes several basic weather and QG diagnostics displays at 1200 UTC 18 December 2009. Examine the `Div(Qvector)` field to gain familiarity with regions of QG forcing for ascent and descent. Additionally, **capture an image of the 700-hPa** `Omega` **field. Where is the geographical location of strongest upward vertical velocity at the 700-hPa level, and what is the peak value?**

b) Now, **open the LMT_3.3 bundle.** *Alert!* Uncheck `Remove all displays & data` in the `Open bundle` dialog before proceeding:

You may have to click OK on an IDV error message or two, but don't worry about them; they are related to this bundle's use of theta (in units of K) as a vertical coordinate.

Three additional LMT_3.3 displays will appear under `Isentropic Displays` in the `Legend`. Colored contours of pressure and wind barbs are shown *on the 300-K isentropic surface.*

Now let's compare the vertical motion computed from isentropic techniques to the QG-diagnosed vertical motion from a). First, ensure that only the 300-K winds and pressure contours are showing to avoid clutter, and examine the pattern of wind and pressure on this isentropic surface (the 300-K surface) at 1200 UTC 18 December 2009. **In what general regions are the winds blowing "uphill" towards lower pressure? Where are they blowing downhill?** Now, in order to check your answer, activate the display of `PresAdv` (pressure advection, $-\vec{V} \cdot \nabla_\theta p$). **Which sign of pressure advection corresponds to ascent?** This is *the rate of change of pressure for air parcels if the flow were adiabatic and the theta surface did not move with time.* It has units of Pa s^{-1}, directly comparable to the `Omega` display. Could you show that mathematically?

Toggle the `Omega` display on and off. **How well does the pattern match the pattern you expect from** `PresAdv`**? How well do the peak values agree quantitatively?** If upward motion drives latent heat release, **how would you expect that to affect the magnitude of actual omega, relative to the adiabatic estimate in** `PresAdv`**? Explain.**

c) **How well do the locations where the isentropic flow is blowing uphill match the regions of QG forcing for ascent, as diagnosed in a) above? Are there any regions where the isentropic method indicates ascent where the QG method did not, or vice versa? Discuss.** *Hint: Think about what vertical level in the atmosphere we are examining here, and consider changing the display to other vertical levels to see how much sensitivity there is to this selection.*

d) Recall that it is the *storm-relative* velocity, rather than the observed wind velocity, that enters into the isentropic vertical motion equation [Eq. (3.12) in MSM]. **How might this affect the diagnosed vertical motion pattern from c)? Discuss.**

Toggle the `Div(Qvector)` display on and off. **How well does the pattern match the pattern you expect from `PresAdv`?**

Toggle the `Omega` display on and off. **How well does the pattern match the pattern you expect from `PresAdv`? How well do the peak values agree quantitatively?**

How does the upward motion field correspond to the cloud field? Activate the IR satellite display under `Isentropic Displays` (note the IR display from LMT_2.2 that has only one time level). Examine the first time (1200 UTC 18 December), and subsequent times (only the `Isentropic Displays` will update with time). **Discuss the extent to which the satellite image matches the isentropic pressure advection. What factors would explain any discrepancies?**

If upward motion drives latent heat release, how would you expect that to affect the quantitative *magnitude* of actual omega, relative to the adiabatic estimate in `PresAdv`? Explain.

3.4. Isentropic Techniques for Analysis and Prediction of Current Weather

The objectives of this exercise are 1) to practice making your own IDV bundle from scratch, and 2) to apply isentropic analysis to a current forecast situation.

First, this exercise walks you through the process of making up an IDV bundle from scratch. This procedure is detailed here because isentropic values and slopes are substantially different across the seasons, so an off-the-shelf pre-made bundle may not give you a clear view without adjustments. Since the adjustments need to be described anyway, this exercise elaborates the entire construction from scratch.

In IDV, on the `Dashboard` go to `Data Choosers`, and `Catalogs`. In the window to the right of the `Catalogs:` label at the top of the page, type http://lmt. meas.ncsu.edu:8443/repository?output=thredds.catalog or http://ramadda.atmos.albany. edu:8080/repository. Under the `data, Isentropic` folder, locate the subfolder `GFS Isentropic Interpolations`. Select the files `Latest gfs Isen` and `Latest gfs211` with a double click or the `Add Source` button. We are ready to plot some current isentropic data.

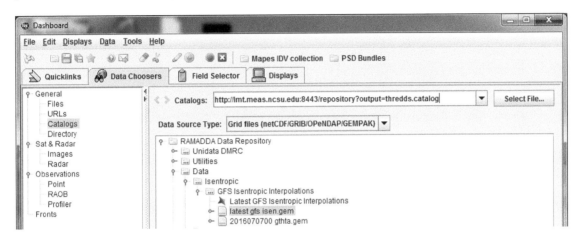

One drawback to the isentropic framework is that for different locations and seasons, one must plot data on different isentropic surfaces. How do you know what surface is "right" for a given forecast scenario? The following activities address this question.

a) We are often interested in diagnosing or predicting precipitation. This technique informs us about vertical air motions, and we know that most atmospheric water vapor resides in the lower troposphere. Thus, our goal is to diagnose and anticipate vertical air motions which take place *in the altitude range where precipitation formation is most prevalent*. Given the above, **what approximate range of pressure levels should bracket the region of interest** on the isentropic surfaces we should plot?

Consider your geographical location and season. If you are conducting your analysis in a warm location or for a warmer time of year, you will need to plot a higher-valued potential temperature isosurface. For colder seasons and more northerly locations, plot a lower value. Recall that surface temperature is roughly equal to potential temperature at low-elevation stations, and that the troposphere is about 30-50K deep in terms of its range of θ values. **Discuss how you chose a potential temperature value for your initial isosurface plot.**

b) The *slope* of isentropic surfaces also varies with season and location. In general, we would want to plot pressure contours on an isentropic surface with an interval as small as 10 hPa for "small slope" situations, and perhaps as large as 50 hPa for situations in which the isentropes slope strongly. In the dashboard, click edit and `change display unit` in order to change the units to millibars from Pascals. **During what time of year would you expect to observe the largest isentropic slopes, and why?**

Now, on the IDV `Dashboard` in the field selector, let's contour the isentropic pressure (pressure @THTA) with a contour interval consistent with the consideration above. Make your selections as shown below and click `Create Display`.

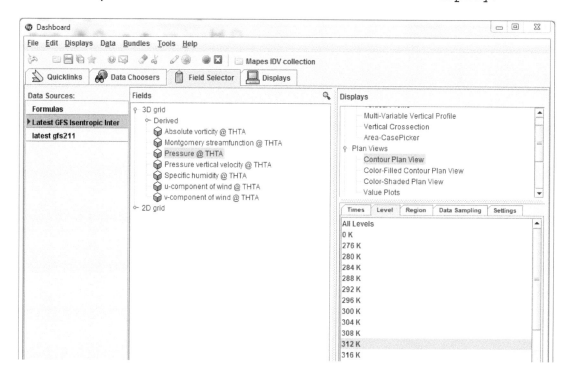

If nothing appears right away, that's okay; we will need to make a few changes because IDV is not designed to handle isentropic coordinates. After the display is created, right click on the new display in the `Legend` and under `Edit`, select `Change Display Unit`, changing from Pa to millibar as shown here:

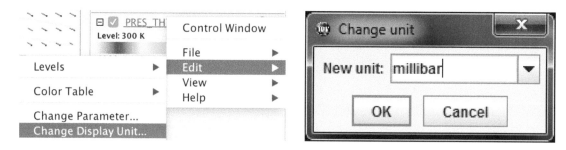

Now click on the `Legend` hyperlink for your `PRES_THTA` display, and adjust the `Contour` properties (interval, range, etc.) as needed to fit the current situation. Change the contour display color to white, and apply some smoothing in order to make the contours less angular.

To provide more precise geographical references, click `Default Background Maps` in the Legend and select `Hi-Res US`:

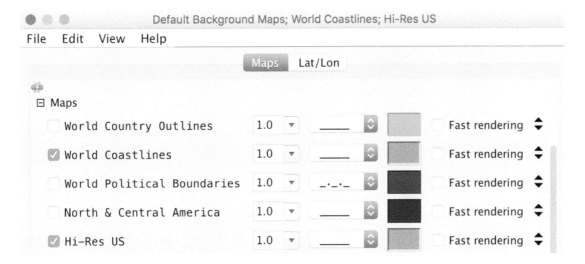

Now, we need to overlay the wind field on our isentropic surface in order to identify regions of isentropic ascent or descent. In the `Field Selector` tab, under the `Derived` list, select `True wind vectors`, and then click on `Create display`:

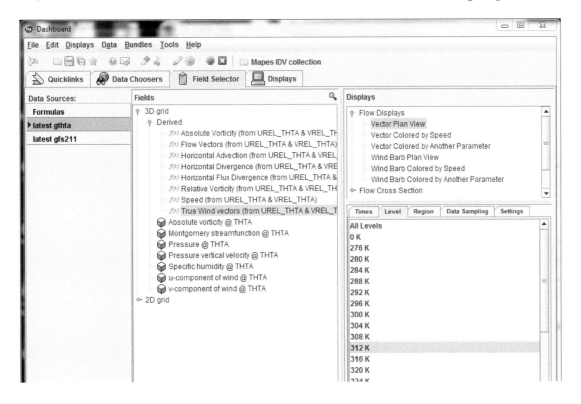

You will probably want to adjust these wind vectors, depending on how strong the flow is for the current example. Example settings are shown below for a light-flow situation:

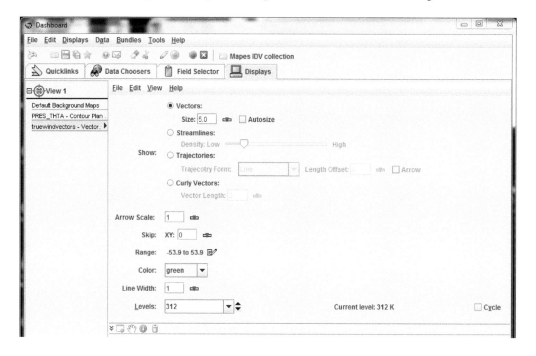

Now, you are ready to examine the forecast.

c) At the initial time, **is there evidence for isentropic ascent anywhere over your location of interest? If so, where, and how can you tell?** Orient yourself—where is the isentrope at a higher altitude, and where is it lower? Use the labels, and adjust settings if you need more of them.

d) Now, step through the forecast sequence, focusing in on the area of interest. **When and where do any prominent areas of isentropic ascent develop?**

Return to the `Field Selector` tab, and add a `Color-Filled Contour Plan View` display of specific humidity or mixing ratio on the selected isentropic surface. Adjust the contours so that shading appears only for values above a useful threshold, like these:

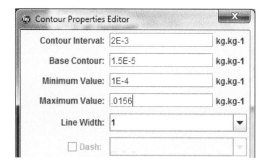

Apply a color scheme, such as the `Mapes GreenHaze` option:

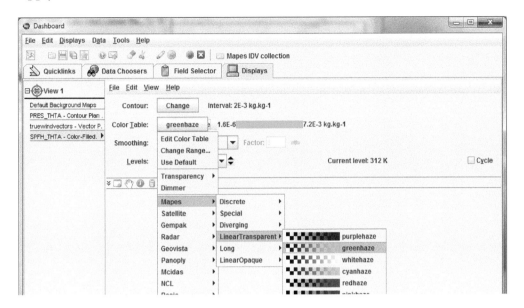

e) Based on the combined information in the water vapor and isentropic flow analysis, **make a forecast of expected cloud cover and precipitation for your location for the next few days. Save and discuss one key image that shows isentropic pressure, wind, and shaded mixing ratio on the selected isentropic surface. For example:**

Basic cloud/precipitation forecast for site in question, with "isentropic justification":

	First 12-h period	Second 12-h period	Third 12-h period
Expected predominant cloud cover (OVC, BKN, SCT, FEW, CLR)			
Expected precipitation (percentage for 24-h period)			
Isentropic reasoning to support forecast			

At this point, right click on the `PRES_THTA` display in the `Legend,` and go to "view," "times," and set this display as the time driver (`Drive times with this display`).

Finally, in order to check our isentropic reasoning against the model precipitation forecast, now return to the `Field Selector,` and select the pressure-coordinate data file `latest_gfs211.gem,` which should correspond to the same model run as the isentropic data file we used. This grid file contains the data in standard isobaric coordinates, but it also includes the model precipitation forecast. From the field selector, go to `2D grid,` and select `Total precipitation 6 hour @none.` Before creating the display, under `Times,` set `Match Time Driver` as indicated below:

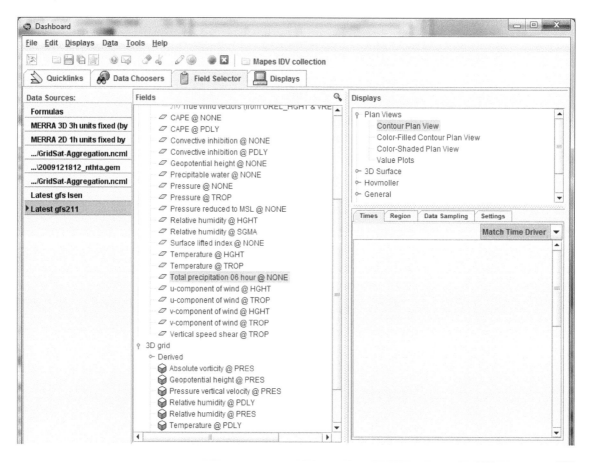

Apply some smoothing to the contours, and recognize that the times will not match the isentropic display exactly, so pay attention to the labels at the bottom of the display window. Edit and change the display unit to inches, and change the contour interval to successive values of `0.01; 0.05; 0.1; 0.25; 0.50` (this will only display these discrete contour values).

f) **To what extent does the model precipitation forecast match your isentropic prediction from e)? Discuss. How much precipitation does the model predict for your specific location of interest?**

g) In the gfs211 isobaric data set, **create a 3D isosurface of the selected isentrope** (`3D > Derived, Potential Temperature (from TMPK_PRES`). Turn off the other displays for clarity, and zoom out for a view of the entire continental United States. Change the angle to look at the isentrope from an oblique 3D perspective. **What features are present that were associated with any regions of strong isentropic ascent or descent?** [Hint: A "feature" could be an upper-level trough or ridge, jet streak, etc.]

Having invested this effort, click `File`, and `Save as` to save this bundle with a descriptive name, something like "*my_LMT3.4_isentropic.xidv*". In the future, when you open this bundle, you can once again view the current day's isentropic datasets.

References

Carlson, T. N., 1980: Airflow through midlatitude cyclones and the comma cloud pattern. *Mon. Wea. Rev.*, 108, 1498–1509.

Kocin, P. J., and L. W. Uccellini, 1990: *Snowstorms along the Northeast Coast of the United States: 1955–1985. Meteor. Monogr.*, Vol. 44, Amer. Meteor. Soc., 280 pp.

4

THE POTENTIAL VORTICITY FRAMEWORK

The focus of chapter 4 is the potential vorticity (PV) framework for weather analysis. This is an advanced topic, and the treatment here is not intended to be comprehensive. Note that chapters 2, 5, and 7 also contain related material on PV.

This chapter includes the following exercises:

4.1. Potential Vorticity and the Tropopause in Synoptic Systems
4.2. Potential Vorticity in a Winter Storm
4.3. Jet Streak PV Interpretation
4.4. Diabatic Processes and PV Evolution: The Diabatic Rossby Vortex

Each exercise in this manual uses these typefaces for clarity:

Normal typeface is used for background information, technical instructions, motivating questions, and learning objectives. **Bold indicates assigned actions and questions that students are expected to respond to in their report.** A `constant width` typeface is used to indicate text that can be found exactly on the IDV software (usually on the `Dashboard` or `Legend` areas).

The word **Optional:** is used to set off suggestions for further explorations.

4.1. Potential Vorticity and the Tropopause in Synoptic Systems

Recall that the Rossby–Ertel form of the potential vorticity (PV) is proportional to the *product* of the static stability and the absolute vorticity. The quasi-geostrophic (QG) approximation to PV is given by the *sum* of the *geostrophic* absolute vorticity and the static stability. Both express the insight that when wind convergence increases vorticity, it also spreads isentropes apart (decreasing static stability), so that the combination is conserved. Specifically, *for adiabatic, frictionless conditions, the value of PV will remain constant following the flow.* This property helps spotlight areas where diabatic processes, such as condensational heating, are affecting the PV distribution. For example, if an isolated lower-tropospheric PV anomaly develops *in situ* (without advecting in from some other region), then we can be sure that a diabatic or frictional process was responsible. This aspect is explored more in the following lesson (4.2).

Why is it useful to know the PV distribution? In part, it is because of another property of PV, *invertibility*. If we know the PV distribution, and provide boundary conditions, the PV anomaly can be "inverted" via iterative numerical solution techniques to recover the associated wind, temperature, and pressure/height fields. Cyclonic PV anomalies are, as one would expect, associated with cyclonic rotation locally, but the associated cyclonic flow field can extend far away from the immediate vicinity of the PV anomaly itself.

The very large static stability in the stratosphere makes potential vorticity there much larger in magnitude than in the troposphere, where the average static stability is much smaller. Therefore, one can define the *dynamic tropopause* as a constant PV surface with a value that typically represents the lower boundary of large-PV air, such as the 1.5- or 2.0-PVU surface.* The pressure of the dynamic tropopause surface indicates its altitude (which can also be characterized by potential temperature θ). Winds on this constant-PV surface indicate its direction of motion, since it is a material surface, a sheet of parcels with the same value of a conserved tracer. In this way, an elegant summary of upper-tropospheric (and lower stratospheric) dynamics can be shown in a single map!

The objectives of this exercise are to compare traditional plots of geopotential height and wind to the PV distribution so that students can relate familiar isobaric charts to the PV framework. By the end of the exercise, students should be able to explain the relation between troughs and ridges in the height field to centers of relatively large or small PV, and to the topography of the dynamic tropopause. We will again utilize the December

---

* 1 PVU = 10^{-6} K kg^{-1} m^2 s^{-1}. In this unit, the s^{-1} is from vorticity, while gravity g converts dθ/dp in K/Pa to K kg^{-1} m^2.

2009 winter storm for this exercise, which will by now be familiar to students from earlier exercises.

Open the bundle LMT_4.1. Initially, only the 250-hPa geopotential height contours and wind barbs are visible, valid at 1200 UTC 18 December 2009, revealing an upper-level trough over the central United States.

Comparison of PV and features in the height field

a) Activate the `PV at 250` displays to show the 250-hPa level PV distribution.* **Examine the relative values of PV in the trough over the central United States, as well as in the regions corresponding to ridges in the geopotential height field located over the western and southeastern United States. Based on this comparison, describe and discuss the correspondence between PV and synoptic-scale features in the height field.**

b) Next, deactivate the `HGHT PRES Contour Plan View` display, and compare the wind barbs to the PV distribution at this level. **Discuss what you see regarding the relation between features in the wind field and the PV, including regions of cyclonic and anticyclonic flow, without the distraction of height contours.**

c) Now, we will consider the surface upon which the PV is equal to 1.5 PVU, the *dynamic tropopause* as discussed in the introduction above. Re-activate the `250 hPa HGHT PRES Contour Plan View` display and turn off the `PV at 250` displays. Activate the `pvor - Isosurface colored by Height` display. **Discuss the correspondence between the altitude of the 1.5-PVU surface and the synoptic-scale trough and ridge features noted previously. In other words, does a higher or lower altitude of the dynamic tropopause correspond to cyclonic flow and a trough in the 250-hPa geopotential height field? Explain.**

d) Let's peek at the lower troposphere: Activate the `1000 hPa HGHT PRES Contour Plan View` display and rotate the display so that you can look "underneath" the dynamic tropopause with a viewpoint from the southeast, similar to that shown below. You will see that a lower-tropospheric cyclone is present over the Gulf Coast:

---

* Recall that PV is computed using a layer in order to represent the static stability $d\theta/dp$. Here, the 250-hPa PV reflects a finite difference of θ across the 200–300-hPa layer.

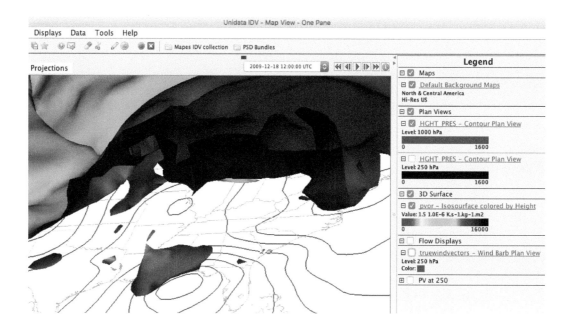

Notice that there are some locations that exhibit "stratospheric" PV values but that are located close to the surface (red-orange colors; e.g., over the Gulf Coast as shown here). **Using your knowledge of the PV conservation principle and the possible diabatic sources for PV, speculate as to the processes responsible for this red lower-tropospheric cyclonic PV maximum, and any others you may see at low altitudes.**

e) Go to a top view, and activate the `IR satellite` display. **What would you look for that would be consistent with the development of the lower-tropospheric PV anomaly you examined in iv. above? Capture an image, and explain how it supports or refutes your speculation from iv.**

4.2. Potential Vorticity in a Winter Storm

As mentioned in exercise 4.1, the conservation property of PV means that for adiabatic, frictionless flow conditions, the value of PV will remain constant following the flow. Usually winter storms involve filaments or fragments of PV from the tropopause-level polar vortex moving across the midlatitudes at upper levels. In some cases, an isolated lower-tropospheric PV anomaly develops *in situ* (i.e., without advecting in from some other region), so that we can be sure that a diabatic or frictional process was responsible. Suppose that condensational heating leads to the generation of a lower-tropospheric cyclonic PV maximum. Balanced "PV thinking" would then allow us to infer the presence of cyclonic flow in the vicinity of that PV anomaly, and we would know that this flow was the dynamical result of the heating. Of course, in order to quantify the effect, performing an actual PV inversion would be necessary.

Before delving too far into the workings of PV, it is useful to examine its relation to the more traditional QG variables presented in chapter 2. Exercise 4.1 began that comparison, and we will extend it here. In this exercise, we examine an event from both QG and PV perspectives to better appreciate the consistency between these frameworks. For instance, in the case of a low-level heating-induced cyclone described above, recall from the QG height-tendency equation (exercise 2.6) that *below a heating maximum* there is a *negative geopotential height tendency*. But at the same time, this lowering of geopotential heights can be viewed as an aspect of the balanced flow around the cyclonic PV maximum created by the vertical heating gradient term in the PV tendency equation.

The objectives and learning outcomes for this lesson are for students to 1) compare interpretations of atmospheric dynamics derived from traditional QG approaches to the PV approach, 2) relate specific physical processes to PV anomalies, and 3) associate atmospheric flow anomalies with specific portions of the PV distribution.

The case we will use for this exercise is a famous winter storm that took place on Presidents' Day in 1979 (18–19 February 1979). This exercise shows data from high-resolution numerical simulations of this case, using reanalysis boundary conditions. Using simulated data allows us to better quantify specific physical processes during this event.

Open the bundle LMT_4.2. Initially, only the 300-hPa geopotential height contours valid 0600 UTC 19 February 1979 are visible, revealing an upper-level trough over the eastern United States.

a) Assessment of QG forcing: Vorticity and thermal advection terms

    i.   **Based only on the 300-hPa height field, where would you expect to observe a cyclonic vorticity maximum?** Activate the box for `absvort - color-filled contour plan view` to check your guess. **Based on this upper-level vorticity pattern, where would you observe QG forcing for midlevel ascent?**

    ii. Next, activate the `PMSL` display to overlay contours of sea level pressure, showing a coastal low as well as a weaker minimum under the upper trough. We can infer the geostrophic temperature advection by looking for regions of veering or backing geostrophic wind with height. **Considering the change in geostrophic wind direction between the surface (based on SLP contours) and upper troposphere, identify regions of warm advection, coinciding with QG forcing for ascent.**

iii. Based on the traditional form of the QG omega equation, consider both *cyclonic vorticity advection increasing with height* and *warm advection* to identify expected regions of QG forcing for ascent. **In light of the implied QG forcing for ascent, and taking into account the likely availability of moisture, where do you expect that precipitation might be falling at this time? Save an image showing the areas of expected precipitation at this time based on this QG reasoning.** One way to annotate a map is with the IDV's `drawing control,` invoked with the pencil ✐ icon in the `Toolbar.` Its use can be learned from the Help menu.

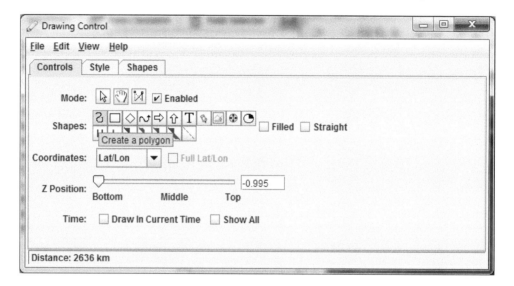

iv. Select the `REFC` box to overlay contours of model-simulated radar reflectivity valid at this time (0600 UTC 19 February 1979). **Do the areas showing precipitation match your expectations from iii. above? Are there regions where you expected to find precipitation but it is absent? Are there areas showing precipitation where you did not expect it from QG reasoning? Discuss the extent to which your overall expectations were met, list at least one area with a discrepancy, and offer an explanation as to the possible cause.**

b) The vorticity and PV viewpoint

The synoptic picture shown so far features a vigorous upper-level trough approaching the U.S. East Coast, with forcing for ascent and widespread precipitation taking place in the mid-Atlantic region. Now, let's examine the vorticity and PV structure accompanying this system.

i. Deactivate all displays except for 300-hPa `HGHT` and `PMSL,` and then activate the `absvort - Isosurface` display of the 25×10^{-5} s^{-1} absolute vorticity isosurface. Rotate and examine this surface. Notice that there are at least two large regions of high vorticity enclosed by this surface: a ragged one near the coastal surface cyclone, and a smoother curved sheet at upper levels in the vicinity of the upper-level trough. **Capture some images showing these features, and write brief captions expressing what you see in each image.**

ii. Now, turn off the `absvort - Isosurface`, and activate the `complete pvor isosurface`. This surface represents the 2.0-PVU surface, which is one way to define the *dynamic tropopause* (DT), the boundary separating high-PV air in the stratosphere from generally smaller tropospheric PV values. Rotate and examine this isosurface. The structure of this surface is complex, but it carries a great deal of information about various physical and dynamical processes. **Describe its major (that is, large-scale) features, capturing images as needed.** If you orient the display so that you are looking a bit southward from above, you will notice a "hollow" extension of the isosurface extending down towards the ground. This feature is known as a "tropopause fold." While looking down into the fold, reactivate the `absvort - Isosurface` vorticity isosurface. **Toggle back and forth between the PV and vorticity isosurfaces. Describe and interpret what you see.**

iii. Given that PV is conserved in the absence of diabatic processes, and that the latent heat associated with precipitation is all far to the east of the tropopause fold, **how did this high PV air come to be observed way down in the lower troposphere?**

iv. Recall that PV is a product of absolute vorticity and static stability, so that high PV values occur wherever *one or both* factors are large. Activate the `Vertical Cross Sections` displays, disable the `isosurface` displays so you can see better, and **capture a view from the west. Which factor is large in the high PV region (the color shading): vorticity or static stability? Is the answer different at different altitudes?** Recall that the vorticity equation has a convergence term on the right-hand side, and that convergence is related to the vertical motions that you can infer from conserved tracers like PV. **How must the large absolute vorticity inside the tropopause fold have developed? Explain, using a sketch of the cross section at times before and during the tropopause folding event seen in the IDV at this time.**

c) Lower-tropospheric PV

Now let's turn our attention to the lower troposphere, in the vicinity of the surface cyclone. Activate the `low-level pvor isosurface`, instead of the `complete pvor isosurface`. Now the isosurface only spans the 1000–600-hPa layer. Activate the `Plan View` display of radar reflectivity (REFC_NONE).

i. **Use your knowledge of PV conservation properties and the meaning of radar reflectivity to offer a possible source for the cyclonic low-level PV at low levels near the mid-Atlantic states of the East Coast.**

ii. For this case, a computation of the latent heating rate due to condensation was made (LATHT PRES). Activate the `LATHT_PRES - Isosurface` display and the corresponding `LATHT_PRES - Color shaded Cross Section` displays. Viewing the region of precipitation from the south side, gradually increase the `Isosurface` value to see how the vertical heating profile varies, or examine a sweep of the cross section through the area. **Does the vertical *gradient* of latent heating rate match the structure of the cyclonic PV features? Explain, based on the PV equation,** referring to MSM material such as Figs. 4.4 and 4.5 if that helps. **Discuss and illustrate with screen captures.**

iii. Notice that there are fairly large cyclonic (positive) values of PV located far to the west and north of the cyclone and upper trough, very near the surface (e.g., from Minnesota northward into Canada). **What diabatic processes, specifically, may be able to explain the presence of these shallow PV features? Use the relationship from ii.**

d) **Optional:** Repeat the above exercise with the **LMT_4.2_MERRA** bundle to see how this situation looks in a global reanalysis, at a coarser resolution. In that case, `prectot` is the indicator of surface precipitation, and *total* heating `dtdttot` replaces the latent heating field used above.

e) **Optional:** Repeat the above exercise with the **LMT_4.2_MERRA_StormOfCentury** bundle to see the displays of c) for another famous case (mentioned in Appendix 2 of the introduction). *This bundle uses the servers at NASA, rather than zipped data. It can therefore be relocated to any other region of the world and any time during 1979–2015.* To relocate, zoom out and shift-rubberband the desired region, then change the `Time Driver` in the `Animation Properties`. The data and horizontal displays will then load in your new region. After loading is complete, you may have to move the locations of the cross sections manually into your new data hypercube.

4.3. Jet Streak PV Interpretation

This exercise is an application of potential vorticity concepts to a jet streak, a topic presented by Cunningham and Keyser (2000). The objective of this exercise is to understand how a jet streak would propagate, from a PV perspective. Consider an isolated, idealized jet streak of the type pictured on the plan-view plot below:

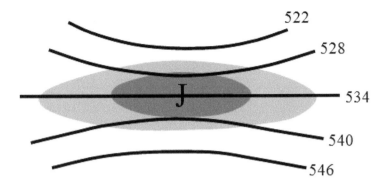

a) **How would one go about representing a jet streak, such as the idealized example above, *in terms of PV?*** In other words, if we accept that the flow in the core of the jet is mostly balanced (generally a very good assumption), then we should be able to attribute it to PV anomalies of either sign. **Where would such anomalies be located in relation to the jet streak? Sketch and label the expected location of PV anomaly center(s) on the diagram above.** Recall that the vorticity represents a component of the PV; revisiting exercise 2.1h may be helpful.

b) Jet streaks are observed to *propagate*; that is, their motion *cannot* be explained purely by advection of wind speed through their core; jet streaks move _____ (faster or slower) than the wind speeds in their core (see textbook for discussion). **But if the jet streak can be attributed to PV anomalies, which *do* move entirely due to advection when the flow is adiabatic and frictionless, then is there a paradox? Reconcile this observation using "PV thinking."**

c) **Is the motion of a jet streak affected by *diabatic* processes? If so, how? Does the flow near the center of a jet streak conserve PV? Discuss.**

4.4. Diabatic Processes and PV Evolution: The Diabatic Rossby Vortex

The following exercise is an application of potential vorticity concepts to a specific situation in which diabatic processes are important. Your task is to utilize PV concepts to predict the behavior and evolution of this particular weather system.

Consider a baroclinic zone, characterized by isentropes that slope upward to the north (consistent with westerly geostrophic wind shear). An isolated zone of ascent has developed at some longitude, leading to clouds and precipitation at the location indicated in the plan-view map and cross section. Assume that the plan-view map corresponds to an altitude of 2 km.

a) Based on your knowledge of PV *conservation* (or lack of conservation: the source or sink terms for "Q" = PV in the equation given below), **in this cross-sectional view, indicate locations of development for any PV maxima or minima.**

b) Based on your knowledge of *PV invertibility*, **sketch the associated horizontal flow associated with this PV feature (or these features) on the plan view diagram.**

c) Based on your answer to b), **how would the *vertical motion* field evolve? Indicate areas of ascent and descent on the plan-view diagram.**

d) Based on c), **in which direction would this *weather system (and associated PV anomalies) move, or propagate,* and why? Consider both the larger-scale ambient flow and the impacts of system-generated flow.** Recall the β effect and barotropic Rossby waves from exercise 1.5, which this exercise generalizes to include condensation effects.

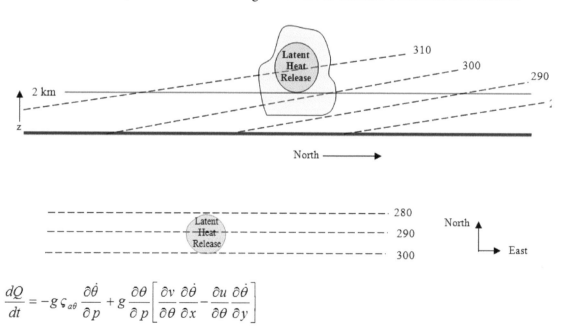

$$\frac{dQ}{dt} = -g\,\varsigma_{a\theta}\,\frac{\partial\dot\theta}{\partial p} + g\,\frac{\partial\theta}{\partial p}\left[\frac{\partial v}{\partial\theta}\frac{\partial\dot\theta}{\partial x} - \frac{\partial u}{\partial\theta}\frac{\partial\dot\theta}{\partial y}\right]$$

Reference

Cunningham, P., and D. Keyser, 2000: Analytical and numerical modelling of jet streaks: Barotropic dynamics. *Quart. J. Roy. Meteor. Soc.* **126,** 3187–3217, doi:10.1002/qj.49712657010.

5

EXTRATROPICAL CYCLONES

Chapter 5 of MSM covers extratropical cyclones, a topic that allows an integration and application of ideas from each of the first four chapters. Cyclones involve several distinct physical and dynamical processes, and a variety of analysis techniques are needed to dissect their structure. Since we have discussed these processes in earlier chapters, the treatment here will not be exhaustive.

This chapter includes the following exercises:

5.1. Cyclone Self-Development Mechanism
5.2. The East Coast Snowstorm of 24–25 January 2000 and Its Forecasts
5.3. The East Coast Snowstorm of 24–25 January 2000: PV Perspective
5.4. Tendencies Comprising the Thermal Budget of a Midlatitude Cyclone

Each exercise in this manual uses these typefaces for clarity:

Normal typeface is used for background information, technical instructions, motivating questions, and learning objectives. **Bold indicates assigned actions and questions that students are expected to respond to in their report.** A `constant width` typeface is used to indicate text that can be found exactly on the IDV software (usually on the `Dashboard` or `Legend` areas).

The word **Optional:** is used to set off suggestions for further explorations.

5.1. Cyclone Self-Development Mechanism

The objective of this short exercise is for students to identify the relative locations of upper- and lower-tropospheric disturbances, and the accompanying temperature advection patterns, associated with the Sutcliffe–Petterssen cyclone self-development mechanism. This material corresponds to chapter 5.3.5 in the MSM text.

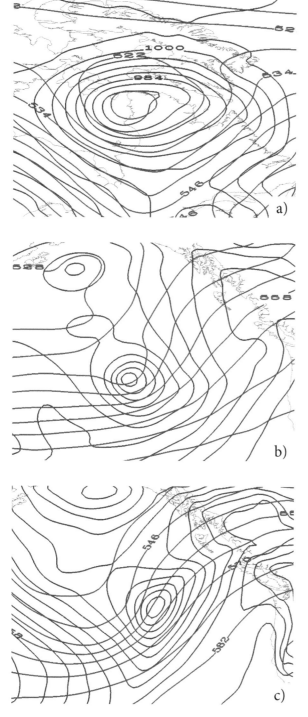

Figure 5.1. Sea level pressure (blue contours, 2- or 4-hPa interval) and 500-hPa geopotential height contours (red, 6-dam interval) for three Northern Hemisphere scenarios. Locations are (a) Hudson's Bay and (b), (c) off the Pacific Northwest coast.

Consider the three plots in Fig. 5.1. Each shows 500-hPa height (red contours, 6-dam interval) and sea level pressure (blue lines, 2- or 4-hPa interval).

a) **Which of these three panels displays the surface cyclone that is in the most favorable situation for the Sutcliffe–Petterssen self-development mechanism to operate? Explain and justify your answer. What aspects of the unfavorable examples make self-development unlikely? What aspect of the most favorable situation makes it conducive to this process? Be specific.**

b) **Sketch sea-level isobars and 500-hPa heights** for a situation in which a *surface cyclone is located to the west of an upper-level trough* embedded in a westerly jet stream. **Is this situation conducive to self-development? Why or why not?**

c) **Does the self-development mechanism apply to an *anti*cyclone, located to the east of an upper-tropospheric ridge? Discuss.**

5.2. The East Coast Snowstorm of 24–25 January 2000 and Its Forecasts

A major winter storm affected the U.S. Mid-Atlantic region on 24–25 January 2000. Preceding the snowstorm, a significant ice storm affected northern Georgia and portions of northwest South Carolina on 22–23 January 2000. Over 500,000 utility customers were without power during and after the storm, with the Atlanta metropolitan area severely affected.

On the heels of the ice storm, a rapidly deepening low pressure center formed along a stationary frontal boundary, and proceeded to blanket much of the East Coast with heavy snowfall on 24 and 25 January 2000. The storm was associated with heavy snow from the Carolinas into New England, with at least five related fatalities reported. At Raleigh-Durham (RDU) airport in North Carolina, the snowfall total from Monday evening the 24th through 2000 UTC on the 25th was 20.3 inches, breaking their old record for a single storm event of 17.9 inches measured on 15–17 February 1902. The snow also fell in a short period of time, with Raleigh-Durham reporting *15 inches in 4 hours* at one point during the early morning hours of 25 January! During the peak of the storm, tropical storm-force winds buffeted the coastal waters of the Carolinas, with Chesapeake Light reporting north-northeasterly winds of 74 mph at 4:00 a.m. on 25 January, while Diamond Shoals and Frying Pan Shoals both reported wind speeds as high as 64 mph.

The storm developed rapidly; Cape Hatteras recorded a pressure fall of 29 hPa in 24 hours, from 1014 hPa at 0600 UTC 24 January to 985 hPa at 0600 UTC 25 January. Numerical guidance from the operational forecast models was poor.

In this exercise, we will explore the following questions: *What signals were available that would have indicated that a major storm system was imminent? Can the poor model forecasts be attributed to physical processes that were misrepresented or not accounted for in the numerical models?*

Today's numerical weather prediction systems have advanced considerably relative to those in use at the time of this event. Further, the concept of ensemble forecasting systems was not yet as widely used at the time of the January 2000 event. Nevertheless, this example is instructive to consider.

a) The model forecast

   Open the LMT_5.2 bundle, and examine the sequence of 500-hPa geopotential height and vorticity from the *NAM model forecast* initialized at 0000 UTC 24 January 2000.

   i. First we will consider the upper-level synoptic features associated with this event. Step through the sequence of 500-hPa height and vorticity for this model forecast. Locate the upper-level trough as it approaches the East Coast. **What aspects of the upper-level trough are favorable for its intensification? What aspects are unfavorable? Discuss and explain**, accounting for the QG processes you learned in chapter 2's exercises.

   ii. Surface features: Now focus attention on the near-surface fields. Uncheck the box for `Upper Level` displays, and activate the `Near Surface` displays of `SLP` (sea level pressure, shaded) and `t2m` (near-surface temperature, contours). Consider the overall synoptic pattern. **Discuss any aspects of the lower-tropospheric synoptic environment that are favorable for the development of a surface cyclone, along with any that are unfavorable.** In this discussion, draw on your knowledge of energy conversion, vorticity dynamics, or QG processes from earlier chapters. **What changes are evident in the sea level pressure field as the upper trough approaches the coast? Why do these changes take place? Discuss the mechanisms at work. Illustrate your discussion with images** as appropriate.

   iii. Vorticity: Now, we will consider the near-surface fields in more detail. Activate the `Near Surface` display of `relvort` (blue contours of 1000-hPa relative vorticity). At the initial time, an elongated zone of enhanced pre-existing vorticity extends from the northeastern Gulf of Mexico northeastward to the waters offshore of the mid-Atlantic region. Two cyclonic centers are seen at the northern end of this vorticity zone. Note that an elongated region of lower sea level pressure roughly corresponds to this band of larger 1000-hPa vorticity (a frontal zone). **Using your knowledge of vorticity, and of the synoptic pattern described above, explain why this synoptic setup is conducive to vorticity generation (cyclogenesis).**

iv. Forecast: Display the 500-hPa height contours again. Step through the forecast sequence. **Write a 1–2-sentence summary of the sequence of events relating to surface cyclogenesis along the East Coast in this NAM model forecast. Capture an image showing ONLY the 500-hPa heights and sea level pressure for the 24-h forecast valid 0000 UTC 25 January. We will use this image for later reference.**

v. Precipitation forecast: Finally, activate the 6-hourly precipitation forecast displays, called `P06M`. Note that the contours will not appear until the display is set to at or beyond the 6-h forecast, because models are not initialized with the precipitation field. **Does this model forecast produce heavy precipitation in the interior sections of the Carolinas, Virginia, and Mid-Atlantic region (e.g., Washington, DC)? Describe the extent to which heavy precipitation falls on these major metropolitan areas in this model forecast.** The display unit is kg m^{-2}, which is numerically equivalent to millimeters of liquid water; 6-h accumulations are the contoured field. For a U.S. audience, you may want to convert to inches.

b) What actually happened

i. Satellite perspective: **Open the LMT_5.2_addIRsat bundle. Alert! Uncheck** `Remove all displays and data` **in the** `Open bundle` **dialog**: you want to `Add to the current window` the new satellite display. This bundle contains irregularly spaced IR satellite imagery corresponding to the period of the model forecast, including the period of heaviest snowfall. Go to the first image in the sequence, for 0600 UTC 24 January 2000. Relate specific cloud features to features in the upper troposphere and near the surface from the model forecast that we examined previously. Specifically, **what upper-level feature is likely associated with the enhanced cloud feature extending from Tennessee southward towards the northern Gulf of Mexico at 0600 UTC 24 January?** To inform your reasoning, activate `HGHT PRES` to display the model 500-hPa heights. Step through the sequence of satellite images. **Describe what happens to the specific cloud band mentioned above (initially centered over Alabama) as the system approaches the East Coast. Does the sequence of satellite images appear to match the model forecast? Explain.**

ii. A later analysis and its forecast loop: **Open the LMT_5.2_addFCST2.zidv bundle. Again, uncheck `Remove all displays and data` in the `Open bundle` dialog.** The new displays will appear under `FCST2` in the `Legend`. Eliminate the satellite imagery display with the trashcan icon, so that the time levels revert to regular 6-hourly time intervals. **Compare the model analysis valid 0000 UTC 25 January (under `FCST2`) to the 24-h forecast you saved in iv. of part a) above, valid at this time** (displays are categorized under `Upper Level` and `Near Surface` in the `Legend`). **Describe the key differences between the analysis valid 0000 UTC 25 January and the 24-h forecast valid at that time, both at the surface and aloft. Capture images for comparison.**

c) **Optional:** How does the event look in a reanalysis made many years later? Open the LMT_5.2_MERRA_1979-2015.zidv bundle. Analyzed fields with the same display conventions used above are shown. **How well does the MERRA reanalysis capture the case above?** Data are available online from NASA hourly from 1979 to 2015. You can reposition the display area with shift-rubberband, and change the Time Driver to display a longer sequence or other dates (other storms), as described in the introduction.

5.3. The East Coast Snowstorm of 24–25 January 2000: PV Perspective

The objective of this exercise is to utilize the PV framework to obtain a more specific understanding of the physical processes associated with the model forecast failure seen in exercise 5.2.

As you may have noticed in the previous exercise, the *analyzed* cyclone from 0000 UTC 25 January was considerably stronger than what was shown in the 24-h forecast valid at that time. Now, our task is to try to understand the physical processes that led to this difference. First, we will look at the NAM's forecast, and then contrast it with what actually happened, via the use of a later analysis dataset.

*Note: The potential vorticity displays in **LMT_5.3_PV1 and LMT_5.3_PV2** only work in IDV 5.4 and above.*

a) **Open the LMT_5.3_PV1 bundle**, a forecast scenario initialized at 0000 UTC 24 January. You will see a pink cloud-like volume rendering of lower-tropospheric (900–600 hPa) PV, along with sea level pressure and 850-hPa wind vectors. A side view will show you that upper-level PV has been omitted in these displays. Step forward from 0000 UTC on the 25th (which is the 24-h forecast). **Is the surface cyclone in the model forecast accompanied by a prominent lower-tropospheric cyclonic PV feature in this 24-h**

forecast? Save an image, zoomed in on the southeastern United States. Recalling that PV is conserved for adiabatic, frictionless flow, **what physical processes might be contributing to the development of this lower-tropospheric cyclonic PV anomaly?**

b) **Open the LMT_5.3_PV2 bundle.** Examine this analysis valid at 0000 UTC 25 January. **Compare and contrast this to the corresponding forecast fields from a) above, and speculate as to what physical processes could have led to the differences that you observe between the forecast and analysis.** As before, **save an image, and juxtapose it with your image from the 24-h forecast valid at the same time [from exercise 5.2. a) iv.].** Given the physical process you identified above, **what meteorological fields or data sources might have been helpful to forecasters during this event?**

5.4. Tendencies Comprising the Thermal Budget of a Midlatitude Cyclone

The exercises above indicated that diabatic heating can be a factor in winter storms, altering the PV fields and thereby the ensuing storm evolution. But in many locations, diabatic effects are weaker in the winter, with little sunshine and limited amounts of water vapor able to condense and release latent heat in the cool-season air masses.* Meanwhile, adiabatic dynamical processes (like advection) are strongest in the winter, because strong gradients are present, and winds are strong.

To gain a quantitative perspective on adiabatic vs. diabatic effects in the evolution of the weather, this exercise examines all the terms in the temperature budget equation for the 10 November 1998 storm that is exhaustively analyzed in chapter 8 of the textbook by Wallace and Hobbs (2006). Our tool will be NASA's Modern-Era Retrospective Analysis (MERRA), with its complete heat budget data, along with some raw satellite data from NOAA to indicate whether the cyclone is accurately analyzed in MERRA.

In a data-assimilating atmosphere model, the heat budget for local tendency of temperature $\partial T/\partial t$ or potential temperature $\partial\theta/\partial t$ can be written schematically as three terms: fluid dynamical tendencies *dyn*, heating and cooling by the model's representations of physical processes *phy*, and "analysis tendencies" *ana* [see exercise 5.4. e) for an explanation of this term].

$$\frac{\partial T}{\partial t} = dyn + phy + ana$$

---

* Over the oceans, cold air masses flowing over warm water can be subject to strong diabatic heating near the surface.

The dynamical tendency *dyn* is the sum of horizontal and vertical advection by the model's wind field. The physical tendency *phy = rad + mst + trb + fri* is the sum of the following:

- Radiative heating *rad = swr + lwr*, the sum of shortwave or solar heating *swr* and long-wave or infrared heating–cooling *lwr*.
- Moist process heating (such as latent heating) *mst* from the model's convection and cloud schemes.
- Turbulent heating *trb* (another name for this would be *sensible heat flux convergence in the vertical*). For instance, in the case of the boundary layer scheme, this *trb* term is how solar heating of the ground is actually felt by the atmosphere. In the MERRA analysis, this also includes turbulent diffusion that takes place above the boundary layer.
- Frictional heating *fri* is the conversion of the kinetic energy of wind into molecular kinetic energy (that is, heat). It is a small term, but we can see how small.

a) **Open the bundle LMT_5.4.** You will see SLP (`sea level pressure`) contours on a shaded `t2m` temperature map. Can you locate the fronts in this midlatitude cyclone? Toggle the `true IR from satellite` display. **Are the position of the cyclone and its fronts fairly well analyzed in the model fields?** Toggle the display of MERRA's `prectot`, total precipitation produced by the model's convection and cloud schemes. **Does the model's rainfall (and condensation heating) match what you would infer from the satellite imagery? Capture an image indicating a discrepancy between the satellite observations and the model's depiction of the storm. What is your assessment of the general quality of the analysis?**

b) Examine the display of the column-integrated *physical* heating rate `dthdt_phy`. (Here the **th** in d**th**dt means potential temperature θ). Toggle the display of MERRA's `prectot` display on and off. **Do some features in these displays correspond to each other? Which category of physical heating corresponds to MERRA's `prectot`?** Loop between the 1500 and 1800 UTC time steps. **What main differences do you see in non-precipitating regions?** Hints: What is the local time of 1800 UTC in this area? Is the sun shining? Can you interpret some features in the 1800 UTC time step, perhaps with the background map as a guide?

c) Examine the display of the column-integrated *dynamical* (advective) tendency, `dthdt_dyn`. **Does it also have features that correspond to the pattern of MERRA's `prectot`? Explain.** Hint: what are the two big terms that mostly cancel in the temperature budget of a saturated updraft, in a precipitating column of the atmosphere?

d) Examine the `Column Integrated Tendencies` display called `phy+dyn`, the sum of the two fields examined above. **Does this field have much correspondence with `MERRA's prectot`? Why or why not, in light of c) above?** To interpret this field, activate the `Z500` contours display. We can think of the ridge–trough–ridge pattern in `Z500` as a *thickness* pattern (proportional to column temperature), with a cool column of air under the `Z500` trough and its cutoff low. Logically, if the total tendency `phy+dyn` is negative east of the trough, the trough will move eastward. If the tendency is negative within the trough, it will deepen. The same applies to the flanking ridges. **Loop between the two time steps. Using this logic, do the features in the net tendency `phy+dyn` explain the time evolution (motion and/or deepening) of the `Z500` trough/cutoff? Explain, with annotated image captures as needed for clarity.**

e) Examine the display of the column-integrated analysis tendency `dthdt_ana`. This "analysis tendency" is the difference between the *observed* (or more precisely, model-analyzed) *time rate of change* at each grid point $\partial\theta/\partial t$, minus the sum (phy+dyn) of the *model's physical and advective* processes that you examined in d) above. It is the *change that must be put in at analysis time to keep the model's time evolution in line with observations*. If the model's dynamics and physics schemes and its data assimilation system were perfect, this field would be exactly 0 everywhere. If the model or its analysis were completely wrong, this field would have magnitudes as large as $\partial\theta/\partial t$, or even larger. **Is this field "large" in magnitude? Compared to what, and why is that the right comparison in light of these considerations? Does it have features that correspond to `MERRA's prectot`? What is your assessment of the general quality of the model's analysis and physics in depicting this storm?**

In the following steps, you will examine cross sections to show vertical structure.

f) Next let's look at *vertical profiles* of the temperature budget terms. Uncheck the box next to `Atm. Column Integrated Tendencies` in the `Legend` to deactivate all those displays. Click the ⊟ symbol next to that, in order to hide those displays and clean up the `Legend`. Click the corresponding ⊞ symbol next to `Vertical Cross Sections` to expose the displays hidden there. Set the time step to 1500 UTC, since the cross sections only have data at that time. **Activate the first two cross section displays: `Temperature` and `Cloudiness`.** Revisit the `MERRA's prectot` display to notice the position of the sections: they cut through the clouds and precipitation near the heart of the cyclone. Click 🗗 to view the scene from the south. **Where in the cyclone do you see low, middle, and high cloudiness?** Does this agree with your general knowledge of midlatitude cyclones? That is, **is the analysis generally trustworthy in its vertical structure?**

In the following steps, you will examine each of the heating rate displays, one by one.

g) Toggle the display for dtdtlwr (the longwave radiative heating rate). **Can you see the expected pattern of 1) heating at cloud bases,** because the upwelling IR absorbed from the warm Earth below exceeds the downward emission from the relatively cooler cloud base, **and 2) cooling to space at cloud tops? Express the approximate magnitude of the heating in K per day, given that the displayed value is in K s⁻¹.** Now do the same examination of dtdtswr (the shortwave radiative heating rate due to absorption of sunlight) and the sum of the two tendencies (dtdtrad). **Which is more similar to the *net* radiative heating** (toggle dtdtrad to see it)**, the shortwave or the longwave?**

h) Examine the turbulent (dtdttrb) and frictional (dtdtfri) heating terms. **Where are they active? Why? About how big are their largest magnitudes in K day⁻¹?**

i) Examine the moist heating (dtdtmst). (Notice that its positive values saturate the color scale, which is the same for all plots, so you can't tell quite how large it gets in the heavy rain region.) **How closely does dtdtmst correspond to the Cloudiness field?** You should realize that the convection scheme in the model can create precipitation and latent heating without making saturated air and 100% cloudiness on the grid scale. **Where does dtdtmst have negative values? How can you understand those negative values in terms of moist processes (related to latent heat)?**

j) Examine the total of all physical processes (dtdttot). **Capture an image, and annotate features on it which are contributed by 1) moist, 2) radiative, and 3) turbulent processes, synthesizing your findings above.**

k) Examine the dynamical tendency (dtdtdyn). Notice that it saturates the color scale everywhere. **Is the heat budget in this midlatitude cyclone primarily adiabatic?** Next, reactivate the dthdt_dyn display under Column Integrated Tendencies, to **show that the vertical integral over column mass is mainly contributed by the lower troposphere**, below and including the 500-hPa altitude of the Z500 display, which can serve as a reference line on your image.

Examine the analysis tendency (dtdtana). **Is it safe to say that it is much smaller than the dominant term (dtdtdyn)? What would you say about its magnitude relative to some physical tendencies? What does this imply in terms of the quality of the model's physics and dynamics processes?**

Optional: To learn more about these terms in the same storm, consider moving the cross sections. All the dtdt cross sections are bound together, while Temperature and Cloudiness are separately located.

Optional: To study the same set of displays for other cases anywhere other than the dateline, use the bundle **LMT_5.4_MERRA_1979-2015**. This large and complex bundle will access data on NASA and NOAA servers, making very many displays, so that the loading will be slow and may be incomplete—but it still may have value even if some parts fail (click OK to any errors). *If you get impatient with the loading, take a stroll while it loads and think of how long it would take to write code to extract these scientific lessons from these datasets!* Eliminating unwanted displays with the trash can icon before relocating your space and time regions will speed up the process. Save the bundle for yourself if you add any value to it.

To relocate the bundle, the `Time Driver` can be adjusted with the `Animation properties` dialog, found under the animation controller's ⓘ button. The displayed area can be adjusted by zooming out to see your desired region, then using a shift + click-and-drag mouse action to select your area (see screenshot below). The bigger your area, the longer it will take, so think carefully and don't zoom out while it is loading. The cross sections will have to be moved manually into the region of your case's data hypercube, either by mouse actions, or by entering new coordinates into their endpoint `Locations` in one of their `Display Controls`, accessed by clicking its hyperlink in the `Legend`. Happy hunting!

Reference

Wallace, J. M., and P. V. Hobbs, 2006: *Atmospheric Science: An Introductory Survey*. 2nd ed. Elsevier Press, 504 pp. (ISBN: 978-0-12-732951-2. Chapter 8 describes this storm in great detail.)

6

FRONTS

The topic of MSM Chapter 6 is fronts. We know that fronts are important, based on our experience with daily weather. What physical processes lead to the formation of fronts? How can we conceptualize the processes that produce important sensible weather in the vicinity of fronts? How can we identify frontal structure using observations or numerical model forecast output? The conditions that accompany frontal passages at a given location can vary widely. Sometimes, a front will be accompanied by severe weather and heavy precipitation. Other times, a strong front may bring large changes in temperature or humidity, but may not be accompanied by any precipitation whatsoever. What factors are responsible for these different outcomes?

Our objectives in this chapter are to answer the preceding questions. Learning outcomes include 1) recognition of the key processes of frontogenesis from analyzing standard observations, and also from reviewing the derivation of the frontogenesis equation, 2) the ability to utilize the frontogenesis function in weather analysis and forecasting, and 3) developing an IDV bundle, from scratch, to plot the frontogenesis function for a real-data case.

This chapter includes the following exercises:

6.1. Simplified Frontogenesis Function: Derivation and Interpretation
6.2. A Frontal Case Study
6.3. Plotting the Frontogenesis Function Using IDV

Each exercise in this manual uses these typefaces for clarity:

Normal typeface is used for background information, technical instructions, motivating questions, and learning objectives. **Bold indicates assigned actions and questions that students are expected to respond to in their report.** A `constant width` typeface is used to indicate text that can be found exactly on the IDV software (usually on the `Dashboard` or `Legend` areas).

The word **Optional:** is used to set off suggestions for further explorations.

6.1. Simplified Frontogenesis Function: Derivation and Interpretation

One way to understand fronts and the processes leading to their development is to examine the *time-tendency of the magnitude of the horizontal gradient of potential temperature,* in a coordinate system that *follows the wind* (a total or Lagrangian derivative*). This quantity $F = d/dt(|\nabla \theta|)$ is known as the *frontogenesis function.*

Because the midlatitude atmosphere remains close to a state of thermal wind balance, changes in the horizontal temperature gradient are accompanied by changes in the vertical shear of the geostrophic flow. In chapter 2, we saw that ageostrophic circulations must arise in order to maintain this thermal wind balance. The upward branch of these circulations can produce important weather impacts. With some modification (called *semi-geostrophy*) for the sub-synoptic length scales (narrowness) of fronts, the general concepts from quasi-geostrophic theory (the disruption of balance by advection or other processes, necessitating the development of ageostrophic circulations to restore or maintain that balance) can help us to understand frontal weather.

The objectives of this lesson are 1) to derive a simplified version of the frontogenesis equation, 2) to perform a scale analysis to evaluate the relative importance of the terms, and 3) to demonstrate a conceptual understanding of the workings of each term.

Following chapter 6.2 of the MSM text, consider a simplified, rotated coordinate system in which the y' axis is always oriented perpendicular to the front, pointing towards the cold side.

a) Using the definition of the Lagrangian derivative, the chain rule from calculus, and the definition of "F" shown below, **derive the simplified version of the frontogenesis**

equation [Eq. (6.2) in the MSM text]. See section 6.2 of the text for additional information about this equation.

$$\text{"F" is} \equiv \frac{d}{dt}\left(-\frac{\partial\theta}{\partial y'}\right) \qquad \frac{d}{dt} = \frac{\partial}{\partial t} + u\frac{\partial}{\partial x'} + v\frac{\partial}{\partial y'} + \omega\frac{\partial}{\partial p}$$

$$F = \underbrace{\left[\frac{\partial\theta}{\partial x'}\left(\frac{\partial u'}{\partial y'}\right)\right]}_{Term\,A} + \underbrace{\left[\frac{\partial\theta}{\partial y'}\left(\frac{\partial v'}{\partial y'}\right)\right]}_{Term\,B} + \underbrace{\left[\frac{\partial\theta}{\partial p}\left(\frac{\partial\omega}{\partial y'}\right)\right]}_{Term\,C} - \underbrace{\left[\frac{\partial}{\partial y'}\left(\frac{d\theta}{dt}\right)\right]}_{Term\,D} \qquad (6.2)$$

b) Perform a *scale analysis* of the terms in this equation for a given meteorological situation or for typical midlatitude frontal situations. **How do typical values of the basic terms in the vicinity of a front compare to those in general synoptic-scale flows?** Note the different characteristic length scales for spatial derivatives in along- and across-front derivatives.

c) For terms A–D, **draw simple sketches (including isentropes, fronts, and wind barbs) of situations in which that term would be negative (*frontolysis*).**

d) Recall that the *Rossby number* $[U/(fL)]$ is a dimensionless measure of the degree of geostrophy of the flow, formed by taking the ratio of the acceleration term to the Coriolis term in the horizontal momentum equation. Using typical values in the vicinity of a front, **develop appropriate Rossby numbers corresponding to the along-front and across-front force balances. Should we expect the quasigeostrophic (QG) equations, developed in chapter 2, to fully describe the dynamics of frontal systems? Discuss.**

6.2. A Frontal Case Study*

During November 2014, cold Arctic air moved southward into the U.S. lower 48 states from Canada. An intense frontal boundary marked the leading edge of this Arctic air mass. Specific objectives for this exercise are 1) to relate basic meteorological fields to the frontogenesis function and some specific physical processes, 2) to relate frontal circulations to **Q** vectors studies in exercise 2.3 and 2.4, and 3) to develop student ability to utilize the frontogenesis function as a tool in frontal analysis and identification.

---

* We gratefully acknowledge Mr. Tyler Croan, Metropolitan State University, Denver, Colorado, for useful suggestions for this exercise.

Our data source for this exercise will be the NAM 212 grid, with approximately 40-km grid spacing, sufficient to represent the shorter cross-front length scale.

Open the `LMT_6.2_part1` **bundle**, showing the NAM model run initialized at 1200 UTC 11 November 2014. Examine the lower-tropospheric synoptic situation at the initial time. The only field initially displayed is sea level pressure (`EMSL NONE`, contoured). Locate the intense Arctic anticyclone centered over Canada, and the relatively weak cyclone centered near the Great Lakes.

a) Sharper curvature in the sea level pressure contours may suggest where the frontal boundaries are likely located in this case. **Sketch where you would expect the cold and warm fronts to be in this example,** based just on the sea level pressure and your knowledge of how fronts tend to align with the pressure field. To do this, open the IDV's `Drawing Control` (click on the pencil icon ✐ along the top toolbar in the Dashboard). Select the `Draw in Current Time` box so the fronts are only shown on the initial time, and adjust the `Z-position` slider to the top so they will remain visible when other displays are turned on. **Draw your fronts, and save an image displaying your subjective frontal analysis for later reference.**

b) Now, check your estimated frontal positions by plotting the near-surface equivalent potential temperature (with display name `theta-e`). Check the boxes to plot the contour or color shaded displays theta-e field. **Was your initial analysis from a) consistent with the theta-e field, or were adjustments necessary? Discuss.**

c) Next, activate the display of `truewindvectors` in order to consider the action of the 975-hPa wind field on the cold front. Deactivate the SLP contours (`EMSL NONE`) to minimize clutter. **Assess the *shearing* and *confluence* terms of the simplified frontogenesis equation** [terms A and B in Eq. (6.2)]. **For the portion of the cold front extending from Texas to southern Illinois, indicate whether each of these terms is frontolytical, frontogenetical, or weak. Justify your answers.**

d) Like most meteorological software, IDV's "Frontogenesis function" includes only the horizontal advection terms [2D shearing and confluence, terms A and B in Eq. (6.2)]. Examining the wind field and 975-hPa equivalent potential temperature isentropes in the vicinity of both the warm and cold fronts, we can see some places where there is strong implied frontogenesis, and others where it is weaker. **Identify one location where you believe, based on your analysis in c), there would be strong frontogenesis. Then, also along either the warm or cold front, identify a location where you believe the frontogenesis is weak. For each location, explain why you selected it.**

e) Now, activate the `Frontogenesis` contour plan view to check your subjective frontal analysis against the frontogenesis function. Deactivate some of the other quantities in order to reduce clutter. **Is the frontogenesis generally strongest on the warm or towards the cold side of the subjectively analyzed frontal boundaries? Why is this the case?** Hint: Examine terms A and B in the frontogenesis function [Eq. (6.2) in MSM] and think about where these quantities are maximized. **Do the locations from d) that you identified as strong and weak frontogenesis match the frontogenesis calculation? Discuss.**

f) Step through the forecast sequence. **How does the frontogenesis field change in the first 24 hours of this forecast sequence, for the period ending 1200 UTC 12 November? Is the frontogenesis stronger or weaker at 1200 UTC 12 November than it was at the initial time (1200 UTC 11 November)? Discuss any changes, and offer an explanation for any significant ones.**

g) Now **open the LMT_6.2_part2 bundle. Alert! Uncheck `Remove all displays and data` in the `Open bundle` dialog** to add these displays to your existing view. This bundle will add **Q** vectors to the display. We saw above that confluent flow near the front produced *positive* values of frontogenesis at this time (confirmed by the contours of `Frontogenesis` at the 975-hPa level). **Do we expect Q vectors to point towards the warm or cold side of the front in these areas? Explain.** (Hint: recall section 6.3 in the MSM text, or exercise 2.3 in this manual). Now, activate the **Q**-vector plot (`qvec`) and zoom in to various locations along the cold and warm fronts to note the correspondence between the frontogenesis function and the orientation of the **Q** vectors (at the 975-hPa level).

   i. **Describe the relative magnitude of the Q vectors relative to the strength of the frontogenesis. Is Q larger where the frontogenesis is stronger?**

   ii. Using your knowledge of what the **Q** vector represents from chapters 6 and 2, **discuss why, conceptually, we might expect the Q-vector magnitude to be related to the strength of the frontogenesis.** Can you see or show why, mathematically?

   iii. **Is the orientation of the Q-vector field consistent with your expectations, based on the arguments presented in section 6.3.1 of the text?**

h) Kinematic treatments of frontogenesis sometimes neglect the effect of the frontal *secondary circulation* on the frontogenesis itself [a *tilting of θ surfaces* that can influence the magnitude of the horizontal temperature gradient, term C on the right side of the frontogenesis equation (6.2) above]. In a more dynamically complete view, this frontal secondary circulation works in the opposite sense to the horizontal frontogenesis (terms A and B, the main "forcing" for the front's development). We also know that the **Q**-vector convergence should be consistent with the upward vertical branch of the frontal circulation. **If we were to view a cross section perpendicular to the cold front in this case, would we expect to see a** *thermally direct* **or** *thermally indirect* **circulation? Would the vertical component of this circulation, the tilting term, act frontogenetically or frontolytically? Explain.**

i) Activate the `Vertical Cross Section` displays, and examine vertical motion `OMEG_PRES` in the front-normal cross section. Only negative values (upward motion) are displayed. Deactivate other plotted parameters in order to more clearly display the cross-section variables, and **capture an image. Does the vertical motion in the cross section match your expectations based on QG theory and the previous discussion?** Step through the first several time periods in order to gain a sense of how the frontal circulation evolves over the first 24 h of this NAM model forecast. **Is this frontal circulation deep in vertical extent and likely to produce heavy precipitation, or shallow? How does the depth and strength of the circulation change with time during the first 24 hours of this forecast sequence?**

j) Finally, consider diabatic effects on the thermal gradient, term D in (6.2). Notice that the *total derivative* dθ/dt is the *diabatic* heating or cooling of air parcels, distinct from all the partial derivatives in (2). **Open the LMT_6.2_part3 bundle, again adding it to your existing displays.** This displays a satellite image valid at 1800 UTC 11 November superimposed on the Rapid Update Cycle (RUC) model analysis valid at that time. Locate the cold front, based on the warm edge of the potential temperature gradient and the 975-hPa height contours. **Is the diabatic term from latent heat release acting frontogenetically or frontolytically in the cold front at this time? Justify your answer. How about in the warm front? Again, discuss and justify your answer.**

6.3. Plotting the Frontogenesis Function Using IDV

The primary objective of this exercise is to illustrate the process of making a frontogenesis display from any gridded data file. A secondary objective is to allow quantitative comparison of IDV-computed frontogenesis values with the scale analysis from section 6.1.

In the IDV `Data Chooser` in the `Dashboard` window, load today's 1200 UTC NAM run, for example, with 80- or 40-km data (this is available in several different catalogs, but the screenshot below illustrates how you might find it in Unidata's IDV Catalog).

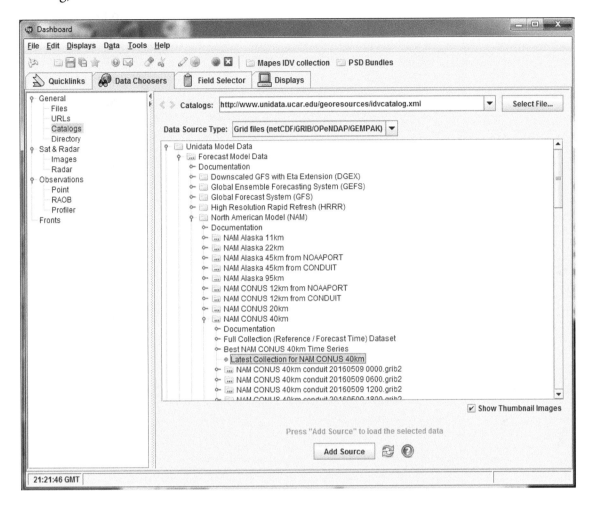

First, **plot the near-surface temperature or potential temperature field as a** `Contour Plan View` **or a** `Color Shaded Plan View` using elementary operations in the `Field Selector` tab. Based on the result, **identify an area of interest that exhibits strong gradients, which could correspond to a front**.

Next, we will plot the frontogenesis function. Remember, the IDV frontogenesis calculation *only* plots the first two right-hand terms in the frontogenesis equation (6.2) above, confluence and shearing, which are just the horizontal, advective parts. It does not plot the last two: the gradient of diabatic heating/cooling, and tilting of θ surfaces, which would require data on heating rate and vertical velocity.

To contour the frontogenesis function, select the following set of pull-down menus from the `Field Selector` tab in the `Dashboard`:

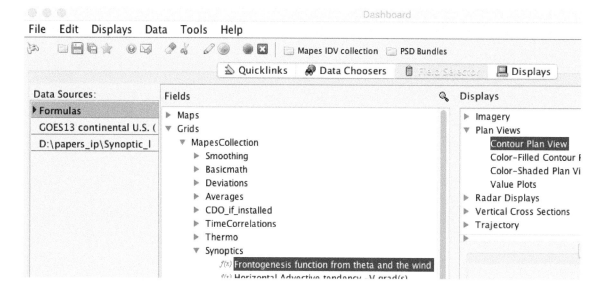

Once you hit `Create Display`, you will be prompted for the fields that are needed for the computation. For the scalar field (S) select `Derived > Potential Temperature` at some lower tropospheric pressure level. For the vector (V) field select `Derived > True Wind Vectors`, which will be drawn from the corresponding vertical level. You will need to choose these from a `3D grid` data set.

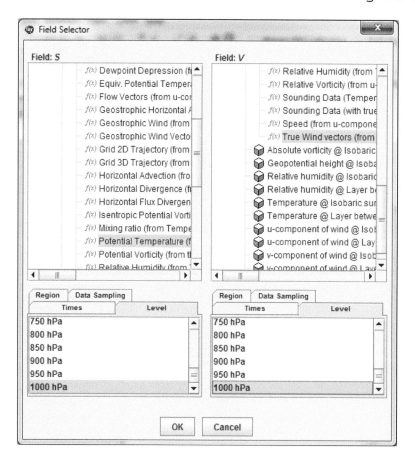

a) **What are the *units* of the frontogenesis function? Does this make intuitive sense?**

b) **Does the order of magnitude of values correspond to your scale analysis from 6.1? Are they in different units? If so, change the units to match.** To do this, click on the frontogenesis function label in the legend and then, under the `Edit` menu, select `Change Display Unit`.

c) **Adjust contour intervals as needed to isolate fronts without too much nonfrontal clutter** (the exact values will be case dependent). **Are there regions that exhibit *frontogenesis* that do not correspond to *traditional fronts*? If so, how can you tell the difference?**

d) **Optional:** Try plotting a 3D isosurface of the `Frontogenesis function` in order to gain a sense of 3D structure. In order to set this up, you will need to use isobaric (3D grids) data. For clarity, consider restricting the range of `Level` for potential temperature and wind (using shift-click on a second value in the `Level` list to select a range), for example 1000 to 800 hPa. What is the three-dimensional structure of the fronts you see? **Do they tend to "lean back" towards the colder air, or are they upright, or do they lean towards warmer air? Why do they seem to have a preferred structural tilt?**

If you are satisfied with the frontogenesis displays you have created, consider saving them as an IDV bundle for future use. To do this, click on `File`, **and** `Save as`**….** Since you selected the `Latest` file from the `Data Choosers - Catalog`, this bundle will always display the current weather.

7

COLD-AIR DAMMING

Note: Previous chapters of the lab manual coincided with the chapter numbers in the MSM text, but beginning with this chapter, they will no longer do so. Chapter 7 in the lab manual corresponds to chapter 8 in the MSM text.

In chapter 2, we saw that quasigeostrophic (QG) vertical motion could be conceptualized as the vertical branch of an ageostrophic response to repair a disruption of thermal-wind balance, with QG omega as the ascending branch of that circulation. Here we ask related questions: *What happens when synoptic-scale flows impinge on topography in a manner that would initially disrupt geostrophic balance? How is the geostrophic adjustment process manifested in the lower troposphere, and what are the impacts of this adjustment on weather and climate near topography?*

An important example of such an adjustment process is cold-air damming (CAD), which we will examine in detail in this chapter. Although we can identify CAD at many different middle- and high-latitude locations around the world, very well-documented examples commonly take place to the east of the Appalachian Mountains in the eastern United States, the focus of the exercises in this chapter.

We will first examine how a winter storm accompanied by CAD differs between a control simulation and a numerical model experiment with the Appalachian Mountains removed. The second exercise will allow us to evaluate some quantitative theoretical parameters that relate to topographic influence on the atmosphere. Together, these exercises allow us to test some of the concepts presented in chapter 8 of the MSM text.

This chapter contains the following:

Each exercise in this manual uses these typefaces for clarity:

Normal typeface is used for background information, technical instructions, motivating questions, and learning objectives. **Bold indicates assigned actions and questions that students are expected to respond to in their report.** A constant width typeface is used to indicate text that can be found exactly on the IDV software (usually on the Dashboard or Legend areas).

The word **Optional:** is used to set off suggestions for further explorations.

7.1. The Appalachians' Role during a Mid-Atlantic Winter Storm

Mid-February of 2003 featured a high-impact winter storm event in the U.S. Mid-Atlantic states. This storm is known in some circles as the "Presidents' Day II" storm, not to be confused with the original Presidents' Day storm of February 1979 (which we examined in exercise 4.2). In Chapter 8 of the MSM text, Figures 8.9–8.13 compare model simulations of the Presidents' Day II event to experimental simulations in which the model terrain was removed, allowing isolation of terrain effects. The objectives of this exercise are to explore these simulations more fully.

Specifically, our objectives are to 1) isolate, identify, and explain the influence of topography on the synoptic-scale and mesoscale pressure, temperature, and wind fields; 2) evaluate topographic influences on the structure, intensity, and orientation of frontal systems during a winter storm event; 3) explain the subsequent influence of the topographically disturbed flow on the path and intensity of a cyclone; and 4) identify the ultimate effect of topography on near-surface sensible weather conditions.

a) Analysis of control simulation

Open the IDV, and **load the LMT_7.1 bundle**.

This bundle contains output from two simulations run with the Weather Research and Forecasting (WRF) Model: one (the Control Run) with and one without topography (No-Topography Run). The simulations here were initialized at

0000 UTC 15 February 2003. You will see sea level pressure, 10-m winds, and 2-m temperature for the `Control Run,` as well as model topography (white contours).

i. Step through the 3-day model `Control Run` simulation sequence and **briefly describe the synoptic evolution (the track and evolution of the main cyclone and any secondary features, and the southward excursions of cold air).**

ii. Now, step forward to 0000 UTC 17 February. **Describe the weather pattern over the southeastern United States in the vicinity of the Appalachian Mountains at this time.** In light of the chapter 6 material on frontal types, how would you classify the front extending from east of North Carolina westward to Georgia, several hundred kilometers southeast of the Appalachians? Justify your answer. **Capture an image for this time from the control simulation, and annotate it to highlight or identify any specific features you refer to. What frontogenetical mechanisms are likely responsible for the location and strength this front? List the mechanisms,** briefly explaining how they influence the location and strength of this front.

iii. Summarize how you think CAD affects the fronts and the sea level pressure pattern you described in i. and ii. **If the Appalachian Mountains were not present, what do you think would be different in the situation you described above?**

b) Comparison with a no-terrain simulation

Now, uncheck the `Control Run` box in the `Legend` to deactivate all the displays from the control simulation from a) above. Check the `No-Topography Run` box to activate similar displays for this experimental simulation. As before, examine the sequence. You may wish to toggle the individual `Control Run` displays on and off for comparison, to assist in the following exercises.

i. Step through the 3-day model simulation sequence, and again **describe the synoptic evolution (the track and evolution of the main low and any secondary features, and the southward excursions of cold air). Is the situation simpler than in the Control Run?** Might it therefore be more conceptually appealing to think of this no-terrain run as the *reference* case, and think of the more realistic run with topographic complications as an "experiment"? This is the sense of the "`effect of topography`" difference fields you will examine below.

ii. Examine the evolution of the surface cyclone in this no-topography model run. **Develop an explanation, using a vorticity perspective, of the differences in** evolution between this simulation and the control run. **Specifically, consider vortex stretching in an air column flowing over a barrier (ignoring frictional effects for the moment).** Also bear in mind that a cold, stable CAD air mass is locked in place east of the actual terrain. Would air flowing over the mountain barrier be able to freely descend in the lee of the mountains with the CAD air mass in place? How would the presence of terrain, and a very cold, stable air mass located immediately to the east of the terrain, affect a cyclone propagating through that region?

iii. Uncheck the `TMPK_HGHT` displays and activate the `t2m difference` display in the `Difference (effect of topography)` legend area. **Where and when are the largest temperature differences caused by topography? How great is the magnitude? When and how does the largest difference feature develop? Specifically for the state of South Carolina, how do sensible weather conditions differ between the control simulation and the no-terrain experiment?**

iv. **Optional:** Make a new SLP difference display, and a wind vector difference display, using `Formulas > Miscellaneous > Simple difference` fields, from the `Field Selector` tab of the `Dashboard` window. As above, make the difference Control Run (`ctl`) minus No-Terrain (`noter`), isolating the `effect of topography`. For scalar fields, adjust the color shading so that the difference display is symmetric around 0, by choosing from `Color Table > Mapes > Diverging` and adjusting `Change Range...` to equal and opposite values like −10 and 10 hPa. At the initial time when there is little difference, the display should be blank (white or transparent, with possible exceptions in areas of high terrain, where the sea level pressure computation would differ between the simulations). **Describe any additional insights you gain from seeing these "anomalous" or difference pressure and wind displays.** For more advanced analysis, under `Data Choosers, Catalogs,` you can connect to the 3D `cad ctl` and `cad noter` model data sets in the LMT chapter 7 area within the Mapes IDV collection, and display many additional fields from these simulations.

v. **Imagine that the Appalachian Mountains are roughly double their observed height. How do you expect that sensible weather would differ in the vicinity of the mountains if that were the case? Explain and discuss, both for this particular weather event, and generally.** How might we go about testing this speculation?

7.2. Analysis and Interpretation of Theoretical CAD Parameters

a) Surface analysis interpretation

We will now examine a CAD event from February 2001, with the objective of evaluating two theoretical parameters using real case data. Specifically, this exercise requires computation of the Froude number and the Rossby radius of deformation, with the goal of enhancing our understanding of the physical processes involved.

First, we will examine the synoptic evolution of this event using a sequence of surface analyses. Open the **LMT_7.2.zidv** bundle. This bundle includes surface and upper-air observations, along with analyses from the NAM model.

i. The first display time shows surface observations and the NAM model analysis valid at 0000 UTC 11 February 2001. At this time, a large Arctic high pressure system is moving across central and eastern North America. The leading edge of the arctic air mass can be identified as a cold front over the western North Atlantic. Activate the 1000-hPa equivalent potential temperature contours (`thetae`) to reveal the gradient on the cold side of this front. Now, step ahead to 1200 UTC 11 February. Consider the orientation of the isobars relative to the Appalachian Mountains over western Virginia, North and South Carolina, and northern Georgia. **Do you see any evidence of ridging in the "cold-air damming region" east of the Appalachian Mountains at this time? What wind direction is evident in surface observations located to the west of the Appalachians, over Tennessee? How about to the east of the Appalachians over North Carolina? Does the wind appear to be in geostrophic balance? (You may wish to zoom in to reveal more surface observations.) Explain your answer.**

ii. Now step ahead to 0000 UTC 12 February. At this time, **what evidence do you see for "adjustment" of the pressure field in the vicinity of the Appalachians? Capture an image, after first identifying any relevant trough or ridge axes using a dashed line (labeled "trough" or "ridge")** using the IDV's `Drawing Tool` under the pencil icon ✏ or some other annotation technique.

iii. Step ahead to the analysis for 1200 UTC 12 February. Examine the surface observations over South and North Carolina. **How many different precipitation types can you identify?** Note that Columbia, South Carolina, is reporting rain with a temperature of 34°F, a temperature that is often cold enough to support snow. **Speculate as to why rain is falling in spite of this cold temperature. Explain how this also relates to the observed precipitation over south-central regions of North Carolina.**

iv. By 0000 UTC 13 February, the CAD event is in the decaying stages, as the parent high has moved offshore and cold advection from the north has weakened in the Carolinas. Note the weak cyclone that has formed east of South Carolina. **Why do you suppose that a cyclone formed there? It didn't seem to move in from anywhere else, but it just formed in place in that region.** Activate the 500-hPa height display (HGHT_PRES). **Is there an upper disturbance to the west of the forming surface low? Why might this region be favorable for cyclogenesis despite the upper air pattern? Explain.**

b) Analysis of the Froude Number, Fr

A profile of potential temperature, computed from the upper-air sounding at Greensboro, North Carolina (KGSO), from 0000 UTC 12 February is shown below in Fig. 7.1. This plot shows potential temperature as a function of height, along with the observed winds. We will use this sounding to estimate first the Froude number, and then the Rossby radius of deformation for this cold-air damming episode.

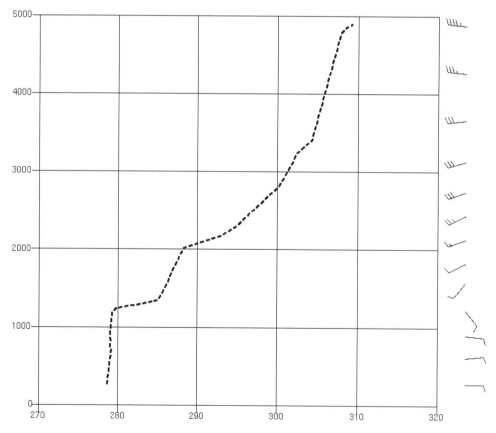

Figure 7.1. Profile of potential temperature and wind as a function of height, taken from station KGSO at 0000 UTC 12 February 2001. For reference, the potential temperature at an altitude of 1241 m is 279.8 K, and at 1341 m it is 285.0 K.

i. First, consider the potential temperature profile provided in Fig. 7.1. **Which portion of the sounding corresponds to the "mixed layer"? Explain.**

ii. An important concept regarding cold air damming is that *large static stability is required*, so that the flow will be blocked by the mountains rather than simply rising up and over them. Mixed layers, such as the one discussed above, are actually layers of very small stability (less stable than usual). If the height of the Appalachian range is roughly 1500 m (a generous estimate), **briefly discuss why we might see CAD in this case, even with a mixed layer near the surface.**

iii. Static stability can be measured by N, the *Brunt-Väisälä frequency*. **Compute N^2 and N including the units the information provided in Fig. 7.1. Show your work.** To do this, use the finite difference form $N^2 \approx (g/\theta_0)\Delta\theta/\Delta z$, where $\Delta\theta$ is the change in θ across a stably stratified layer, Δz is the depth over which $\Delta\theta$ is evaluated, and θ_0 is the potential temperature of the air being lifted (we can use the value in the mixed layer in this case). Take $\Delta\theta$ to be the change in potential temperature between the surface and the top of the inversion layer, and estimate the corresponding Δz.

iv. **Now, compute the Froude number Fr, assuming a barrier height H of 1500 m. What are the "units" of Fr?** Recall that the *Froude number* measures the ratio of the kinetic energy of the wind to the potential energy required to lift air to the top of a barrier. It can be written $F_r = U/NH$, where U is the magnitude of the wind component normal to the mountain barrier and N is your value from iii. above. The value of U can be estimated by averaging the lowest three winds plotted in the KGSO sounding of Fig. 7.1. Wind direction changes little with height within the mixed layer, but then veers rapidly within the inversion layer. Since the mixed-layer wind is blowing at roughly a 70° angle to the Appalachian axis, multiplying the sine of this angle with the averaged mixed-layer wind speed of 5.8 m s^{-1} gives a cross-barrier flow component of roughly 5 m s^{-1}. So we can **use $U = 5$ m s^{-1}.** Clearly there is considerable uncertainty or slop in these values!

v. Laboratory experiments with fluids in tanks show that flow is effectively blocked if Fr < 0.5. **Given this, and the value computed above, would you expect flow blocking in this event? Does this seem to be consistent with observations? Discuss.**

c) Analysis of Rossby radius of deformation

The Rossby radius of deformation L_r is the *e*-folding (or exponential decay) distance scale for geostrophic flow along the edge of a cold air mass. For instance, if a dam that had held back a layer of cold air were suddenly removed, gravity would cause the wall of cold air to initially surge forward with a speed c, related to how cold and deep the layer is. But then after Earth rotated significantly (on a time scale of about f^{-1}), the surging flow would be turned into the along-front direction by the Coriolis force. Geostrophic balance would then be established, with the Coriolis force on the along-edge wind balancing the gravity force that makes the cold air want to slump further. The approximate distance covered by the rushing cold air before the Coriolis force stops it would be (speed × time), $L_r = c/f$.

We will use L_r to compute an approximate theoretical width of the cold dome trapped by the mountains. Specifically, L_r is given by $L_r \approx \sqrt{g'H}/f$, where $g' \approx g\Delta\theta/\theta_0$ and other symbols have their usual meanings. You can see that the numerator is the speed c of a gravity current (the wall of cold air described above) of depth H.

i. **Compute L_r, showing your work, including the units, at a precision appropriate to our crude estimates of H and $\Delta\theta$.** Recall that f = $2\Omega\sin(\varphi)$, where $\Omega = 2\pi$ radians/ (1 day) is the Earth's rotation rate. KGSO is located at latitude 36°N.

ii. **Does this value seem consistent with the structure you observe on the surface analyses? For scale, North Carolina's northern border is roughly 500 km wide.**

iii. **If the maximum cold-dome depth is 1500 m, what would you expect the depth to be at a distance L_r away from the barrier? How could we observe if this was the case? Discuss the implications of the exponentially decaying cold-dome depth for forecasting the precipitation type (rain vs. snow) during winter events.**

8

WINTER PRECIPITATION PROCESSES AND PREDICTION

The subject of this chapter is winter storms, corresponding to chapter 9 in the MSM textbook. Specifically, the focus of this chapter is winter precipitation, including the thermodynamic, dynamical, and microphysical processes that affect the temperature profile during wintry precipitation episodes. We have already analyzed data from a major winter storm in chapter 5, but here we will focus on physical and thermodynamic processes, and on techniques to predict winter precipitation.

This chapter includes the following exercises:

8.1. Physical Process and Sounding Interpretation Review
8.2. February 2003 East Coast Storm Case Study

Each exercise in this manual uses these typefaces for clarity:

Normal typeface is used for background information, technical instructions, motivating questions, and learning objectives. **Bold indicates assigned actions and questions that students are expected to respond to in their report.** A `constant width` typeface is used to indicate text that can be found exactly on the IDV software (usually on the `Dashboard` or `Legend` areas).

The word **Optional:** is used to set off suggestions for further explorations.

8.1. Physical Process and Sounding Interpretation Review

Students will be asked to draw on previous knowledge on cloud and precipitation formation processes in this chapter; perhaps this material was covered in a thermodynamics or physical meteorology course.

The objectives of this lesson are 1) to review some aspects of precipitation formation, and 2) to provide a vocabulary list that will be drawn upon in subsequent lessons in this chapter. For additional review information, a useful resource is the COMET MetEd module, located at http://www.meted.ucar.edu/norlat/snow/preciptype/.

a) **Summarize the Bergeron–Findeisen (cold cloud) precipitation mechanism. If you were viewing a plot of a sounding that was launched through a cloud in which the Bergeron process was active, what characteristics would be evident, and why?**

b) **Define the following cloud physics terms in 1 or 2 sentences, in your own words:**

- **Riming**
- **Aggregation**
- **Collision–coalescence**
- **Freezing nucleus**
- **Contact nucleation**
- **Deposition**
- **Supercooled cloud**
- **Mixed-phase cloud**
- **Accretion**

c) Now we will examine factors affecting the thermal profile, seeking evidence in sounding data to aid interpretations. Utilize the images shown below to answer the following questions:

   i. Examine the Greensboro, North Carolina (KGSO), rawinsonde temperature and dew point profiles shown below (Figs. 8.1–8.3), from a winter weather event that took place in December 2002. For the first sounding (Fig. 8.1), **list all of the *physical processes* you can think of that would be affecting the temperature profile in this situation. For each of the processes you list, indicate *specific evidence* in this sounding supporting your argument that the mechanism is at work.** You should be able to identify at least 7 distinct processes based on what is shown in this sounding. Consider energy absorbed or released by phase changes of water substances, as well as adiabatic and synoptic-scale processes.

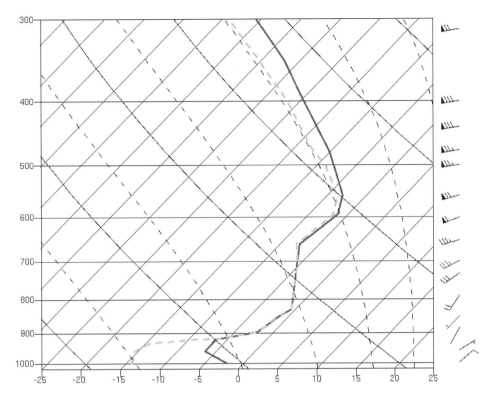

Figure 8.1. Sounding from KGSO from 1800 UTC 4 Dec 2002. Special launch in light of anticipated winter weather event. Temperature (red solid) and dewpoint temperature (green dashed) in skew-*T*, log-*p* format.

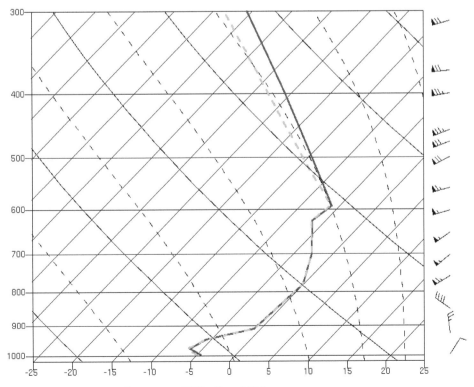

Figure 8.2. Sounding KGSO from 0000 UTC 5 Dec 2002.

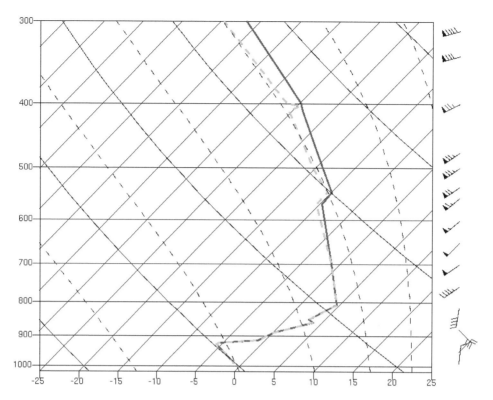

Figure 8.3. Sounding from KGSO from 0600 UTC 5 Dec 2002. Special launch in light of anticipated winter weather event.

ii. Now consider the second Greensboro sounding (Fig. 8.2). **Would different processes be active in this profile relative to that shown in Fig. 8.1? List which processes would be active in this sounding that were not active in the first sounding, and describe the evidence you see in the soundings that support your argument, and explain. Then, list any processes that were active in the first sounding that are not likely to be in the second, and again discuss evidence and explain your answer.**

iii. **Repeat ii. above, but comparing the third KGSO sounding (Fig. 8.3) to the sounding shown in Fig. 8.2.**

8.2. February 2003 East Coast Storm Case Study

Open the LMT_8.2 bundle. It is the largest bundle of this manual (150 MB), so it may take a few minutes to download and render.

The objectives of this exercise are to provide a three-dimensional depiction of the numerically simulated hydrometeor distribution during a high-impact winter storm event. We will utilize output from a mesoscale numerical model for this purpose.

Learning outcomes include 1) evaluating and contrasting the vertical temperature profile during snow, ice pellets, freezing rain, and rain, 2) interpreting changes in hydrometeor type during a complex winter storm, and 3) explaining how precipitation type may be influenced by cold-air damming.

a) In exercise 7.1, we examined the role of terrain during a major wintertime CAD event in the Appalachian region, the so-called Presidents' Day II event. Now, we will utilize this same event (and the control model simulation) to consider the winter precipitation aspects of the case. If the bundle has loaded correctly, the color-filled model-simulated radar reflectivity field should appear (only at times beyond the initial time), superimposed with the 0°C isotherm at the 2-m level. **Loop through the times, and note the geographical regions where, based on these parameters alone, wintry (possibly frozen) precipitation would be expected, provided that this model simulation is accurate.**

b) Step to the time 0000 UTC 17 February 2003, noting the location of the precipitation shield. Now, turn off the plot of radar simulated reflectivity (REFD PRES - Color-Filled Contour), and activate the 0°C Celsius temperature isosurface (T P-Isosurface). This allows visualization of the three-dimensional structure of the freezing level. Rotate the display to examine the structure of this isosurface. **Describe what you see, and discuss the implications of this for expected precipitation type at the surface. Is the structure of the freezing level isosurface unusual? If so, in what way? Discuss.** If you wish, change the color or transparency of this surface in order to better assess its structure.

c) Next, activate the isosurface of snow mixing ratio (r snow p - isosurface). **Evaluate the relative position of the freezing level isosurface and the snow mixing ratio isosurface. Is there any snow located in above-freezing air (beneath the freezing level)? What happens in locations where the freezing isosurface is "folded"? Does snow re-form in the surface-based sub-freezing layer? Why or why not? Discuss and explain.**

d) Now, activate the rain water isosurface (r rain p - isosurface), and deactivate the snow and freezing level isosurfaces (or, toggle them on and off as needed to best visualize what is happening). **Are there locations where rain is present with sub-freezing temperatures?** Finally, activate the graupel isosurface (r graup p - isosurface). **Describe the complex combination of hydrometeor types present in the Mid-Atlantic region during this event. Capture an image that illustrates the relative positions of the different precipitation types.** Consider altering the transparency of these isosurfaces as needed to reveal areas of different, or mixed, precipitation types.

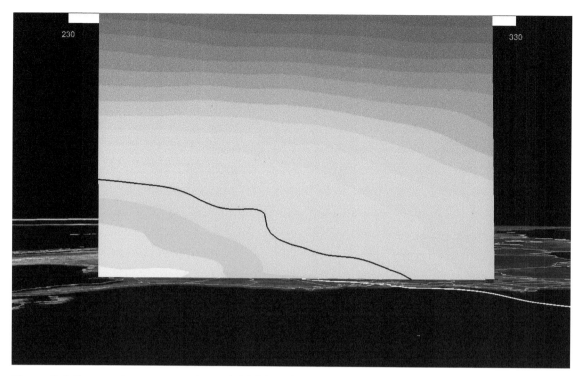

Figure 8.4. Temperature cross section at initial time (0000 UTC 15 Feb 2003).

e) We will now utilize a cross-sectional view in order to examine the hydrometeor distribution more clearly. Consider the location of central Virginia, where the near-surface air temperature is below freezing, but there is warm air aloft at 0000 UTC 17 February. **Deactivate the** `3D Surface / isosurface` **displays, and activate the** `Vertical Cross Section` **displays**, which initially include color-shaded temperature and the 0°C isotherm (black contour). There are also sections available for rain, snow, and graupel mixing ratio. Maneuver the section so that you are looking westward from a perspective off the U.S. East Coast, and return to the initial time; you should see something resembling what is pictured in Fig. 8.4.

Step through the sequence and watch the evolution of the freezing level. Stop at 0000 UTC 17 February. Then, activate the cross section contours of rain water, snow, and graupel mixing ratio. Note that the contour intervals are not the same for all of these quantities.

Discuss the distribution of rain, snow, and graupel in this model cross section relative to the freezing level at 0000 UTC 17 February. What type or types of precipitation are reaching the surface within the cold-air damming region extending from Maryland, central Virginia, North Carolina, and southward to upstate South Carolina and northeastern Georgia?

ACKNOWLEDGMENTS

Many individuals and organizations helped make this manual possible. First and foremost, we acknowledge the critical role of the Unidata program, for their pioneering work in data standards and distribution, for creating and supporting the Integrated Data Viewer (IDV), and for facilitating community uptake. We also gratefully acknowledge the essential role played by the National Science Foundation, which has supplied core funding to the Unidata program over its nearly 40 years of existence. NSF has additionally funded Unidata's Equipment Grant program; funds from this program have been a catalyst in spurring use of the IDV and the data catalog services on which our lab exercises depend. Specifically, we thank individuals at Unidata including Jeff Weber, Yuan Ho, Julien Chastang, and Mohan Ramamurthy, as well as former employees Don Murray and Jeff McWhirter (whose company, Geode Systems, continues to develop the RAMADDA repository software that hosts the Mapes IDV Collection). The American Meteorological Society, specifically Sarah Jane Shangraw and Beth Dayton, provided steady and (very) patient leadership during the multi-year period during which the manual was developed. We also appreciate the copy editing done by Kristin Gilbert, technical editing done by Hollis Baguskas, and the marketing support provided by the AMS. Without the national organizations that gather and freely share observational and numerical modeling data, including NOAA, NASA, and NCAR, data-rich exercises like these would be impossible. The IDV bundles in this manual have benefitted from numerous inputs from students and faculty alike, beyond our ability to comprehensively thank. However, contributions from Prof. Jim Steenburgh, University of Utah, and Mr. Tyler Croan, formerly of Metropolitan State University in Denver, are especially acknowledged. To all of our students, who have patiently tested and provided feedback on earlier versions of the draft and IDV bundles, we thank you for the key role you played in improving the product to this point.

Gary Lackmann wishes to acknowledge the support of his wife of over 25 years, Jen, and his two patient daughters, Sandy and Grace, who tolerate his workaholic tendencies and help to bring balance and inspiration to his life.

Kevin Tyle would like to reiterate the critical role of Unidata; his experiences, first as an attendee at an IDV workshop in 2008 at Plymouth State on up to participating on Unidata's governing and steering committees jump-started his professional development in ways that he never could have imagined less than a decade ago. Kevin also thanks his husband and partner of over 20 years, Bryan LaVigne, for all of his love and support, and for putting up with all the weather and data activities that continue to distract Kevin!